Kaleidoscope: Housing & Living

(1949—2024)

Special Committee for Human Settlements, Chinese Society for Sustainable Development
China National Engineering Research Center for Human Settlements

CHINA CITY PRESS

图书在版编目（CIP）数据

中国人居印象75年：1949-2024 = Kaleidoscope：Housing & Living (1949-2024) / 中国可持续发展研究会人居环境专业委员会，国家住宅与居住环境工程技术研究中心主编. -- 北京：中国城市出版社，2024. 10.

ISBN 978-7-5074-3774-4

Ⅰ. X21

中国国家版本馆CIP数据核字第20245XV403号

责任编辑：宋　凯　毕凤鸣

责任校对：芦欣甜

Kaleidoscope: Housing & Living（1949-2024）

Special Committee for Human Settlements, Chinese Society for Sustainable Development

China National Engineering Research Center for Human Settlements

*

中国城市出版社出版、发行（北京海淀三里河路9号）

各地新华书店、建筑书店经销

华之逸品书装设计制版

建工社（河北）印刷有限公司印刷

*

开本：787毫米×960毫米　1/16　印张：18½　字数：328千字

2024年10月第一版　　2024年10月第一次印刷

定价：**58.00**元

ISBN 978-7-5074-3774-4

（904790）

Kaleidoscope: Housing & Living (1949—2024)

Plan:

Zhang Xiaotong

Chief Editor:

Special Committee for Human Settlements, Chinese Society for Sustainable Development

China National Engineering Research Center for Human Settlements

Executive Editor:

Li Jie

Interview Compilation:

Gao Xiuxiu, Li Yang, Li Yijue, Wu Honglei

Translator:

Fu Jin

Our Impressions on the Changes of Housing Conditions in China

What impressed me the most were thefolk houses of the Va nationality.
The most intuitive impression is undoubtedly the local natural scenery.
Yunnan's species are even more complex.
The local snacks are very diverse and showcase ethic characteristics.
By stepping out of the ivory tower of school and immersing yourself in local life, you will truly observe the unique charm of this borderland town.
Overall, the biggest thing I feel here is a sense of peace.

An Qi, Liu Rui, Pei Jingyi, Yang Yao, Chen Ziyan, Kang Xue'er, Li Jingyi
Student volunteers from China Agricultural University for teaching assistance

I often think that Ushenzhao's pioneering path of desert management and grassland building over half a century ago was absolutely correct—it is a path towards sustainable ecological development. My choice was right; I joined the Communist Party when I was twenty, and for 66 years since then, I have been a Party member without letting down either the Party or the people throughout my life. It has been worth it!

Baori Ledai
Female, Mongolian ethnicity, born in the 1930s, originally from Ushen County, Ordos City, Inner Mongolia Autonomous Region. She has served as the Secretary of the CPC Inner Mongolia Autonomous Region Committee. She was a distinguished member of the 9th, 10th, and 11th Central Committees of the CPC and had also been part of the Standing Committees of the National People's Congress during its 4th and 5th sessions. In recognition of her contributions, she was honored with the National Three-Eight Red Flag Holder Award in 1960.

This improvement signifies more than just a transformation in our living environment; it represents a leap in quality of life and an enhancement in our overall well-being. The moment we moved into our new dwelling was filled with joy and contentment for the entire family as it is not only a place where we live but also a sanctuary resting our soles upon it. We are eagerly looking forward to spending more wonderful time here while creating precious memories in our new abode.

Cao Rui

Female, born in the 2000s, Han nationality, originally from Heze City in Shandong Province, and currently a student at Guangxi Normal University.

After discussion, our family decided to choose the land lot to build a new house for our parents; after all, we Chinese adhere to the traditional life creed as "Returning to hometown, the root and destination of life." The twilight of our lives shall be serenely in the embrace of our ancestral land, both for our parents and ourselves. With support from incentive policies of the government, our dream of having a new house will soon become a reality. The whole family is exceptionally happy about it. What a promising future for us!

Chen Jintai

Male, born in the 1970s, Han nationality, from Zunyi City in Guizhou Province. A literature enthusiast. His works have been published on party's newspapers and new media websites and won many awards in various competitions.

Nyingchi receives abundant rainfall and boasts rich forest resources. The minority folk houses are constructed based on local conditions using indigenous materials, showcasing distinctive ethnic characteristics.

Shanzeng Ciren

Male, born in the 1980s, Tibetan ethnicity, originally from Linzhi in the Xizang Autonomous Region, currently working as a cadre in a state-owned enterprise.

Our family eventually settled inan apartment at the central square of the city. By residinghere, we are able to enjoy an urban lifestyle with convenient transportation and shopping facilities while participating in various group activities like local citizens do. Underneath warm sunlight, many elderly people gather together where we frequentlyengage in discussions on ways to maintain a healthy body and mind.

Dong Xuewen, Jin Yanxia

Born in the 1940s, Han nationality. The couple were originally from Ningxia Hui Autonomous Region. Xuewen Dong, the husband, has served as both a barefoot doctor and a teacher at a rural middle school.

Looking back on all these residences, despite the satisfactory residential conditions and community environment we now enjoy, I still hope to have the opportunity to live in our old ancestral house in my hometown. It is not merely about owning a detached property, but rather about fulfilling the dream of embracing a pastoral farming lifestyle: engaging in work at sunrise and finding solace at sunset, while savoring a tranquil, poetic and traditional countryside.

Duan Meichun

Male, born in the 1980s, Han nationality, a native of Chongqing City, an associate professor of Southwest University

No matter what kind of construction is planned in the village, a Villager Representative Meeting will be launched to solicit opinions from villagers. For example, we have prepared four optional plans for the construction of residential buildings; we would adjust the plans according to villagers' opinions; and finally, every villager will vote to make the decision. Even though such efforts may not achieve 100% satisfaction, we have tried our best to obtain as much consent as possible.

Fan Zhenxi

Male, born in the 1960s, Han nationality, a native of Chengde City, Hebei Province, and was then the Party branch secretary of Zhoutaizi Village, Zhangbaiwan Town, Luanping County, Chengde City.

From my early years as an employee until retirement spanning three decades from late 70s to early 10s, people's life standards have remarkably improved since China's reform and opening up. Throughout these years, I have been honored to take part in numerous wedding ceremonies and visited thenewlywed families. In some way, I could consider myself as aneyewitness to the transformative changes brought forth by this epoch.

Guo Lianjun

Male, born in the 1950s, Han nationality, a native of Harbin City, Heilongjiang Province, and a retired cadre of a state-owned enterprise.

Over the years, I have lived in various places. Among them, the housing experience in Shanwei City of Guangdong Province has left the deepest impression on me. From living in tiled-roof houses to tin houses and residential buildings, I witnessed a significant transformation of the local living environment. However, what remains most memorable are the years I spent living in my grandma's cottage in Hunan Province. It always brings back childhood memories of carefree days spent happily rolling around even in desolate fields.

He Meiting

Female, born in the 1980s, Han nationality and originally from Hunan Province. Sheused to work in both Hunan and Guangdong provinces before engaging in operating the family's B&B hotel business.

I use the livingroom as my own office, where I sit down at my desk and work in the morning until lunch. Before lunch, I like having a walk around the building for half an hour. Thestreets that surround the building are pedestrians, which make it more comfortable to walk or cycling. Also, some Chinese from other areas like to come to our community and play chess, ping pong or practice in the exercise machines. We are very lucky to have all these public facilities around us.

José Manuel Ruiz Guerrero

Male, born in the 1980s, a Spanish architect, arrived in Beijing, in 2012.

For the future development direction of our hometown, my daughter believes it should encompass a comprehensive solution involving tourism, special agriculture, and forestry. Meanwhile, I believe that in order to achieve prosperity, we must prioritize road construction. The establishment and improvement of infrastructure serves as the foundation for any future development. Therefore, it is evident that our greatest efforts should be directed towards designing, planning, renovating, and maintaining of the regional environment while also addressing environmental problems. We believe that in the future, our hometown will become a livable rural area with both a beautiful natural environment and an enchanting humanistic atmosphere.

Kang Bin'an

Male, born in the 1970s, Tujia ethnicity, a native of Enshi City, Hubei Province, an employee of a company

From cramped and narrow school district apartments to spacious and bright commercial apartments, from chaotic and disorganized community to tidy and pleasant living spaces, every step of the transformation has witnessed a significant improvement in the living environment of Harbin City. This is not just a change in physical structures but rather a profound enhancement in living quality and a sense of belonging.

Chen Haiyan

Female, born in the 1970s, Han nationality, from Harbin City, Heilongjiang Province, a staff member in the management department of a central financial enterprise.

Every city has its own unique features and special memories. An increasing number of people are now aware of the importance of preserving urban memory. Improving people's living environment is a long-term strategy, not an issue that can be addressed in 10 or 20 years; instead, it mayrequire 50 or even 100 years to achieve significant results.

Li Hongqin

Male, born in the 1960s, Han nationality. He is a native of Qingdao City, Shandong Province, and has retired from his job.

"Houses are meant for living in, not speculating." If houses were relieved of their excess baggage and simply served as shelters providing peaceful rests from wind and rain, they would become simpler and purer.

Li Pengfei

Male, born in the 1980s, Han nationality. He wasoriginally from Chenzhou City, Hunan Province, and currently resides in Shenzhen City, Guangdong Province as an employee of a company.

Looking back, there used to be dust floating on the city roads.In contrast, at first glance now, you would notice a clean and orderly city with green belts, zebra crossings and isolation lines, creating a livable and pleasant environment.

Yang Jixia

Female, born in the 1970s, Han nationality, a native of Wuhu City, Anhui Province. She works as a freelancer.

My mother's house has also been renovated in recent years,featuring exquisitely carved beams and a bright sunlit room that make it beautiful and grand. The family can bask together in the sunlight or engage in physical exercises, making our lives happier every day. Each stage of life brings its own happiness, and contentment leads to long-lasting joy.

Ma Zhanrong, Han Ping

Born in the 1980s, Hui ethnicity,.Thecouple are natives of Xining City, Qinghai Province. They own a restaurant.

From a historical perspective, the continuous pursuit of infrastructure upgrades is deeply ingrained in the mindset of Chinese people. It's no wonder we have earned the nickname "Infrastructure Giants". The construction of new cities, districts, roads, and bridges all reflect our nation's aspiration for a better life from a specific standpoint.

Niu Cunqi

Male, born in the 1970s, Han nationality, a native of Anyang City, Henan Province, an employee of a state-owned enterprise.

Dazhai Village has been committed to the development of tourism for poverty alleviation for more than 20 years, witnessing earth-shaking changes in terms of roads and other infrastructure, sightseeingconvenience, accommodation capacity, and villagers' awareness and knowledge. Now everyone understands that the terrace landscapes are their "Iron Rice Bowl", and the villagers' awareness of protecting the terrace landscapes has also been promoted.

Pan Baoyu

Male, born in the 1970s, Yao ethnicity, a native of Guilin City, Guangxi Zhuang

Autonomous Region. He was the former Chairman of Village Committee and Secretary of Party Branch Committee of Dazhai Village, Multinational Autonomous County of Longsheng, Guilin City, Guangxi Zhuang Autonomous Region.

Looking back on the past, whether driven by personal passion or the needs of our country, I have embarked on a fulfilling journey from being a young student to becoming a builder, then an architectural designer, and finally a mentor helping train young architects. I sincerely hope that younger generations will achieve even greater accomplishments in the future, just as the ancient Chinese proverb goes: "Originating from blue but surpassing it with bluish green", meaning that worthy disciples excel their masters.

Qing Zhishan

Male, born in the 1950s, Han nationality, a native of Jiangxi Province, the former vice president of Guangxi Blue Sky Technology Co., Ltd.

These changes not only make our lives more convenient and comfortable, but also transcend material enhancement to provide spiritual satisfaction and self-confidence, thereby elevating our quality of life and sense of happiness. We used to live in poverty; however, through education and the support of government policies, we now have more opportunities and choices to enrich our lives.

Qin Jiangxia

Female, born in the 1980s, Tujia ethnicity, a native of Tongren City, Guizhou Province, an employee of a private enterprise.

Over the course of four decades, spanning from adolescence to middle age, it appears as though I have never truly departed from the campus environment. I have witnessed the transformations across various campuses in several cities. Presently, there exists no inclination within me to relocate away from this cherished environment, for I eagerly anticipate the dawn of a new digital era that will bring about a fresh appearance for the campus environment.

Wang Changliu

Male, born in the 1980s, Han nationality, a native of Haikou City, Hainan Province, an associate professor of Southwest Minzu University

Elevators go up and down in high-rise mansions,
Paved roads connect all the ways around.
One-yuan RMB is charged for each bus ride,
Making it affordable and convenient.
A civilized city with courtesy,
This is my beautiful hometown.
Nutritious rice and noodles, delicious fish and meat,
All are served on dining tables.
Stable earnings secure our food and clothing.
Progressing with the epochal trends of social development,
What a thriving prospect!
Thanksgiving dedicated to my dear motherland,
Happy times bring prosperity to this country and peace to our people.

Wang Guangzheng

Male, born in the 1960s, Han nationality, a native of Jincheng City, Shanxi Province, a retired worker.

Standing in the spacious and bright living room, looking out of the window at the vibrant cityscape, my heart is filled with a blend of emotions. Along the way, I have witnessed personal struggles unfolding and observed remarkable changes in the living environment bestowed by this era.

Wu Xiangyang

Male, born in the 1970s, Han nationality, a native of Xinyang City, Henan Province, a scientific researcher.

In the Spring Festival of 2024, I returned to my hometown and noticed significant changes, particularly in the village environment. Over the past two or three years, cement roads and street lights have been constructed in front of each household's courtyard, greatly improving traffic conditions. Every day, venders drive electric tricycles or vans to sell a wide variety of goods to villagers. If you need any thing, simply greet the vendors and they will bring it right to your doorstep; there is no need to go to the town center anymore. Additionally, shared electric vehicles are available for rent throughout the village, and taxis can now be taken here, making transportation to town much more convenient than before.

Yang Yueguan

Female, born in the 1970s, Han nationality, a native of Qujing City, Yunnan Province. She used to be a full-time mother and later held a job.

Cities embody social development while countrysides serve as the roots of our culture. If we can strike a balance between the advantages of both settings by creating more career prospects in rural areas and facilitating more young people's return to their hometowns, it will undoubtedly contribute to rural revitalization, youth development and cultural inheritance within our country. With nationwide implementation of social modernization, I hope that this day can become a reality soon!

Zhai Jiahui

Female, born in the 00s, Han nationality, a native of Puyang City, Henan Province, and currently a student of Minzu University of China

In the past 40 years of my life, from rural to urban areas, and then to modern metropolises; from mud huts to brick houses, and eventually to today's high-rise buildings; from wooden furniture to household electrical appliances, and now a shift to smart home appliances, all these revolutionary changes symbolize our national progress.

Zhang Ting

Female, born in the 1980s, Han nationality, a native of Xinzhou, Shanxi Province. She currently resides in Kashgar Prefecture, Xinjiang Uygur Autonomous Region as a freelancer.

Convenient transportation promises a prosperous future. From mud to cement and then asphalt, the transformation of roads in my hometown holds our sentimental memories; it is also a microcosm of the changes in people's lives.

Zhou Chenhao

Male, born in the 00s, Han nationality, originally from Jianyang City in Sichuan Province, a student of Minzu University of China.

I truly enjoy living in this apartment, probably because I have become afraid of

loneliness as I grow older. I appreciate the courtyard-like layout, which creates a warm and intimate atmosphere.

Zhu Lixin

Female, born in the 1950s, Han nationality, a native of Beijing, a retired teacher.

"Wow, how splendid your home is!" Ever since I started showcasing my home on Wechat Moments, I have frequently received such compliments. Everyone harbors their own dream of an impeccable residence, and for me, it took a complete span of twenty years to ultimately manifestrealize my long-awaited dream home.

Zhu Qingqing

Female, Han nationality, a native of Anji County, Zhejiang Province, a local government official.

Editor's Remarks

"May the mountains and rivers remain undisturbed, while human settlements are appropriately managed." The living environment serves not only as a foundation for material life but also as a source of spiritual connection. Since the establishment of the People's Republic of China, every brick and tile has borne the weight of history and dreams of its people alongside each city's growth. Over the past 75 years, we have witnessed remarkable changes from rudimentary conditions to prosperity; from monolithic appearances to diversity; from meeting basic needs to fulfilling higher aspirations. Each turning poin tembodies the spirit of our times, reflects an artful way of living, and also crystallizes human wisdom.

Kaleidoscope: Housing & Living (1949-2024) has invited thirty representatives from diverse backgrounds to share their personal perspectives on the living environment. From cherished memories passed down by elderly people to aspirations held by the younger generation, as well as observations made by international friends, these voices intertwine like ver sesofa rhythmic poetry of our times. They offer us glimpses into moments that have been buried by history and evoke deep-rooted emotions tied to the land. We express our heartfelt gratitude towards these ordinary individuals whose stories serve as mirrors reflecting both the imprints left behind by time and the evolution of human settlements. Whether it is within rapidly developing cities or during periods when rural areas undergo transformations, each voice plays an indispensable role in this era's Kaleidoscope.

"Kaleidoscope: Housing & Living" series has been regularly updated every decade since its publication, aiming to provide a comprehensive review of historical changes in Chinese living environments. As members of the Special Committee for Human Settlements of the Chinese Society for Sustainable Development, we are actively involved in standard-setting, technology integration, engineering demonstration, and capacity building related to human settlement development. Through our interactions with individuals from varying backgrounds regarding "human settlements", we have recognized that as society progresses economically and people's living standards improve significantly over time, the quality of human settlements undergoes rapid development. Therefore, we have adjusted the frequency of updating the series to every five years in order to promptly reflect and analyze these changes. This adjustment aims to closely align with reality and capture the latest trends in transforming the living environment while providing readers with a more vivid picture of Chinese human settlements.

Kaleidoscope: Housing & Living (1949-2024) is not only a gift to our motherland for its 75th anniversary, but also a passionate review of China's history in developing human settlements. It also reflects our hopes and longings for an auspicious future. Instead of marking the end of our journey, this series signifies the commencement of a new one. Let us embark on this fresh voyage together, employing our wisdom and courage to create a new chapter in the development of living environments during our era.

Editors: Xiaotong Zhang, Jie Li, and Xiuxiu Gao

August, 2024

CONTENTS

Impressions of the Habitat Environment in Zhenkang County

by An Qi, Liu Rui, Pei Jingyi, Yang Yao, Chen Ziyan, Kang Xue'er, Li Jingyi

"What impressed me the most were the folk houses of the Va nationality."

—by An Qi

During my one-year stay in Zhenkang County, I visited numerous places to experience the settlements belonging to different ethnic groups. What impressed me the most were the folk houses of the Va nationality. The Va people reside in mountainous areas along the Lancang River, characterized by a subtropical climate featuring abundant rainfall and high humidity. Therefore, people have to spend half of their days enveloped by clouds and fog year after year. Consequently, local architecture prioritizes designs that are resistant to rain and dampness as its primary features.

The dwellings inhabited by the Va people are typically constructed atop gentle hills amidst expansive mountains. Traditional dwellings feature "four walls rooted in the ground", meaning that three logs with long branches like forks are used as columns and beams, while flat thin wooden strips serve as rafters covered with rows of straws arranged in advance and tied up with rattans. In these dwellings, the Va people maintain a lifestyle of working at sunrise and resting at sunset; despite their slow pace of life, they live with a happy and contented mindset.

"The most intuitive impression is undoubtedly the local natural scenery."

—by Liu Rui

The most intuitive impression is undoubtedly the local natural scenery, exactly as described in books; it's truly beautiful. Unlike the typical four-distinctive-season climate of northern China, it seems like spring resides here all year round, and your eyes are always filled with visions of green trees that never turn yellow or lose their leaves while brilliant flowers never wither away. Whenever you look up, the sky is always azure blue without any haze or dust storms. Such scenery is rarely experienced by people living in big cities. Even a randomly taken photo would capture a wonderful landscape. However, my only regret is perhaps the absence of snow that can refresh the world with its whiteness and purity—which I consider both a gain and loss.

"Yunnan's species are even more complex."

—by Pei Jingyi

The species in Yunnan Province are even more complex. Whether they are plants or animals, there are many types that I have never seen before. A visit to the local farmer's market can be amazing and rewarding, leaving you curious and asking "what is this, and what is that?" The diversity of animals is further reflected in insects due tothe hot and humid weather in the region. In addition to various types of bugs such as cockroaches and flying ants, numerous unknown species exist.

"The local snacks are very diverse and showcase ethic characteristics."

—by Yang Yao

In terms of diet, the local snacks are very diverse and showcase ethnic characteristics. Sour mangoes, sour papayas... You will find that many snacks here have a sour taste. The local rice noodles also possess this flavor, and sour dips will be served even for barbecues. This is what people call the "Dai ethnic flavor". Initially, many of us were not accustomed to this taste, but after trying it several times, we gradually embraced and eventually fell in love with it. Additionally, I believe that the beef here is cooked exceptionally deliciously. The meat is prepared neither too dry to be chewed like firewood nor too fatty to be greasy; when you taste the beef, you can even smell a subtle hint of milk fragrance. I must say that the local beef is the most delectable one I have ever tasted so far. Whenever I

feel fatigued or in a bad mood after hard work, I go out and enjoy a bowl of rice-cake thread with braised beef or indulge in a beef hot pot, and all my troubles and fatigue vanish instantly.

"By stepping out of the ivory tower of school and immersing yourself in local life, you will truly observe the unique charm of this borderland town."

—by Chen Ziyan

In the new town area, there are broad two-way six-lane roads lined with neat street trees. It also houses government departments, institutions, banks, commercial establishments, restaurants, as well as medium and large supermarkets, offering a dazzling array of amenities. The campus where we work as volunteer teachers is located near the mountains alongside the river. Upon entering the school gate, right next to the main road, you will find Zijin Pavilion (Zijin means "Scholar's Robe"), a sundial fountain named "Time Treasuring", an exhibition hall showcasing the school's history, a library, basketball courts, a flag-raising square, and most importantly, teaching buildings dedicated for middle school and high school educations respectively. Extending inwardly at the end of the main road, there is a T-junction that separates the living area from the teaching area on its opposite side. On both sides of this road stand rows of classic sayings depicted from the renowned

ancient teachings found in "the Four Books and Five Classics". Overall, it is a well-designed modern new town with a fully equipped high school.

The People's Square is located opposite the county government, surrounded by trees and rivers, and features a symmetrical design. Even before entering the square, you can hear the rhythms of "Square Dances" resonating one after another. Upon entering the square, you will notice the gentle sounds of the three-stringed harp as well as spinning tops being hit. These two elements represent Zhenkang's traditional culture, known as "Ashuser" (meaning "singing competition"), as well as its traditional spinning-top activity. Ashuser and spinning tops both reflect its ancient culture, characterized by local features of Zhenkang—one focuses on cultural aspects while the other emphasizes the use of force, representing softness and rigidity respectively.They operate harmoniously to exude the true beauty of traditional ethnic culture.

As a volunteer teacher coming to Zhenkang , I am not surprised to experience its wonderful and unique natural beauty; however, I am unexpectedly amazed by its advanced infrastructure. Regarding the cultural environment, the most attractive aspect is the emergence of a new form of culture resulting from the integration of multi-ethnic groups.

"Overall, the biggest thing I feel here is a sense of peace."

—by Kang Xue'er

I arrived here on a lightly rainy afternoon. Local teachers took us on a ride in a unique bus, similar to a golf cart, and we toured around half of the county. It costs two yuan per person and anyone can simply wave to stop it for a ride. During our bus tour, we were provided with some general knowledge about this county. In fact, Zhenkang can be divided into two parts: the new town and the old town. The new town is home to various dense business including clothing industry, catering business, supermarkets as well as government departments, a central square, a passenger bus station, and most banks. These businesses are strategically located through careful urban planning with combined functions. In contrast, the old town appears more haphazardly mixed together with residential areas interspersed among commercial areas and schools. Additionally, some ethnic minority stockades are also scattered across the old town and have become independent zones.

The community where I resided is called Boarderland Homestead, which belongs to a public housing community provided by the local government. It is located within the scope of the new town and is very close to the school and government institutions. The community is adjacent to a park with a lake, equipped with walking trails, pavilions and fitness equipment. There are black swans and ducks in the lake. Every day, I would walk along the shaded lakeside trails from home to school. Occasionally, when I returned home after evening self-study at school, I could see people sitting on the shore fishing; then I would slow down my pace and lower my voice as I passed by.

Home Deep in the Desert

by Baori Ledai

My hometown is Ushenzhao Town in Ushen County, Ordos City (Inner Mongolia Autonomous Region). It is located in the heart of the Maowusu Desert. "Maowusu" means "bad water", indicating the presence of parasites and poor water quality. If horses, sheep, and people drank it, they would get sick. In such a desert heartland plagued by sandstorm and lacking access to safe drinking water sources, the living environment was extremely harsh. However, after more than sixty years of persistent efforts in controlling desertification by multiple generations, the Maowusu Desert has transformed into a green oasis where herders' lives have greatly improved.

When I was young, herders moved from place to place following the growth seasons of grass. However, it was particularly difficult in the Maowusu Desert where 70% of the area consisted of dunes. At that time, herders would often say, "Upon stepping out of the door, a white-sand ridge spreading in front; raising a few black goats and wearing tattered sheepskin coats, we live in a collapsing bungalow named Bengbeng shaped like a Mongolian yurt." The herders would weave branches of salix mongolica into the framework of such a bungalow and cover it with felt rugs to build a simple "Bengbeng" house for temporary living.

Sandstorms were common in the desert. After each sandstorm, these bungalows got buried while black goats ran onto the roofs. Herders had to look for another place to settle down. In the desert, herders often ran out of food and relied on various wild grasses and seeds as sustenance to fill their bellies. For example, edible wild herbs could be dug up and boiled in hot water before being mixed with yogurt as food. There was also a desert plant called Artemisia arenaria whose seeds could be processed and chopped up for consumption. In autumn, sea-buckthorn berries became delicious fruits.

In the 1950s, herders no longer relied solely on a nomadic life, and each family was assigned a specific area of grassland. They built sturdy "Bengbeng bungalows" using branches of salix mongolica as the framework and rammed mud as the walls. The roofs were covered with dry grass, and the doors were made of either dry grass or branches of orsalix mongolica. Wealthier families constructed sturdy and durable Bengbeng bungalows, resembling a type of cave dwellings. After settling down, the herders no longer relied solely on livestock farming; instead, they began to grow crops as well, which greatly improved their lives.

A Bengbeng Bungalow

However, the harsh environment still posed numerous challenges to their daily lives. For example, neighboring households had to share a ground-level well. When sandstorms occurred, people would cover the well with a cowhide and place a large stone on top. However, during particularly sever sandstorms, the well often became buried, making it difficult to locate its opening afterwards. In such cases, obtaining water became a major problem, and herders had to travel long distances to fetch it.

Every household resided on their own grassland, living in scattered vast areas. After the establishment of P. R. China, the village set up a mutual aid cooperative to organize herders and assist each other with tasks that required intensive labor, such as killing sheep for winter storage—which was indeed a critical project. The cooperative would mobilize households to help with this task. Additionally, the Bengbeng bungalows were often buried by sandstorms; thus, the cooperative would also coordinate efforts to clear away sand and excavate these houses. The formation of the mutual aid cooperative helped bring the herders closer together and strengthened their bonds.

Due to the extremely poor family conditions, I was adopted and raised by my foster mother when I was nine years old. I had to take care of all the household chores, such as herding sheep, milking cows, caring for lambs, and taking care of my sick foster mother. I worked diligently and endured hardships without complaint while always remaining optimistic and cheerful. Therefore, both my foster mother and local herders were fond of me.

When I turned twelve years old, Ushenzhao was liberated and a mutual-aid society was established. I actively participated in its activities. During that time, I studied both on my own and through evening school aimed at eradicating illiteracy among herders. By tracing characters on the dunes with reference to books, even though unable to read them yet, I could write them all.

Later on, a teacher came to teach us so that I could become proficient in reading and writing. In fact, I was the top student in the class. Moreover, I played a key role in the mutual-aid society as well as both junior cooperative and senior cooperative. At sixteen years old, I was elected as vice director of the junior cooperative; while at eighteen years old, I became vice director of the senior cooperative and also enrolled in the Communist Youth League (CYL). At the age of twenty, I had the honor of joining the CPC (Communist Party of China), meanwhile serving as deputy village head alongside holding secretary position within local CYL branch.

I was born and raised in the desert, so I had a deep understanding of the challenges faced by herders. The vast desert of Ushenzhao was in desperate need of afforestation. Fixing the sands could preserve the grass that provides food for sheep and further for people. In the past, herders used to carry Artemisia arenaria and salix mongolica from far away for cooking as well as keeping warm. Since these plants could grow in the desert, an idea occurred to me that we could plant them on the dunes to create a green oasis.

At that time, Chairman Mao called for "afforestation of the motherland" which was a great encouragement to me. I led a youth brigade formed by 61 young men and women from the village and local CYL branch, expecting to grow Artemisia arenaria and salix mongolica on the dunes. But things were not as smooth as we thought. A sandstorm almost blew away all the seedlings. Once, we planted 30 *mu* (approximately 20,000 sqm) of seedlings and only three survived.

Some people began to mock us, while others lost hope. However, what I saw was that three seedlings had survived. As long as there were surviving seedlings, there was hope. This time the number was three, and in the future, it could be three hundred, three thousand, or even thirty thousand. Throughout the planting process, despite having seedlings blown away and replanting repeatedly, we also constantly learned from failures and gained experiences. Since it was impossible to grow trees or grasses on dunes directly, we planted them at the foot of the dunes where water and shade could be more easily contained. We also used "ankle braces" to stabilize dunes from floating away and then covered them with trees and grasses like "green jackets".

Sand Control Project

(Photos sources: Wechat Official Accounts including Warm News, Erdos Forestry& Grassland Publication)

With support from both the Party and government leadership, I guided the Ushenzhao's herders in persistent efforts to combat desertification while expanding grasslands as much as possible. As a result, Ushenzhao gained national recognition as an exemplary model for other pasturing regions seeking inspiration from our achievements.We dispelled superstitions about "Alashan grass"—formerly considered sacred but revealed as toxic Achnatherum inebrians—and successfully removed it from our lands. In addition to efficiently planting sandy shrubs and salix mongolica, we also discovered that drought-resistant poplars and willows planted on dunes had a relatively high survival rate, stabilizing the dunes while providing fodder for cattle and sheep. Eventually, these trees formed a "3D airborne pasture" (translator's note: it refers to a feedlot system model that integrates forestry, grassland, and livestock farming). Once the greening areas reached a certain scale, we started cultivating our own seedlings instead of relying on transporting them from elsewhere. Our work truly brought hope to herdsmen.

We gradually accumulated experience through trial and error, witnessing the gradual greening of the endless desert. Over the course of more than ten years, we had planted over 200,000 *mu* (133 sqkm) of trees and 40,000 *mu* (27 sqkm) of grasslands in the desert. We also closed off 120,000 *mu* (80 sqkm) of grasslands for regeneration by banning grazing and improved an additional 80,000 *mu* (53 sqkm) of pastures. I once made a vow to "plant grass and trees in the desert; to collect wood and other materials from the desert", which I believed had been fulfilled. However, the hardships and difficulties involved were beyond imagination.

We had to fight not only against harsh environment, but also against superstitious beliefs and the ridicule of some people. Furthermore, the "Cultural Revolution" also had a negative impacted on me. Thanks to the warmth and care given to me by local herders who understood us, and more importantly, thanks to our Premier Zhou Enlai, who became aware of our efforts in sands control at that time. His trust and encouragement granted me a new lease on life and allowed me to fully devote myself to sandstorm control work. Later on, we also received care and support from leaders at all levels from the county, autonomous region, and state. Fortunately, we did not disappoint them. In the course of practicing sandstorm control, we constantly summarized scientific methods such as "Combination Plantation of Shrubs, Trees, and Grasses", "Shoes and Hats Wearing", "Front Shield with Rear Pull" and "Enclosed Pastures". These methods were successfully promoted nationwide and further drew attention from the World Organization for Combating Desertification.

The "Enclosed Pasture" practice is a pioneering effort of Ushenzhao Town, which can be briefly described as households organizing enclosed pastures where herdsmen graze their livestock, plant trees and grasses, and protect their designated areas. These enclosed pastures encircle the sandy lands one after another, gradually expanding green areas. The expansion started from the village and extended to the township, county and district. This form has continued until now, evolving diverse forms with changes in production and lifestyle.

It has evolved from simple grassland enclosure to combined grassland-forest construction with feedlot management and sand control; from limited small-scale ones to various sizes of large, medium, and small-scale ones; from addressing only winter and spring forage shortage to planned rotational grazing for collective management; from enclosing natural grasslands only to enclosing dunes and shifting sand areas using comprehensive measures; from providing fodder solely livestock industry development to a unique "Enclosed Pasture Economy" involving comprehensive construction and multi-business operations.

By the 1980s, desertification control had achieved good results, greatly improving herders' lives. In the 2000s, substantial ecological and economic benefits were attained, leading to increased state investment. Currently, herders build their houses on grasslands along with shelters for sheep and cattle and hay sheds. All necessary infrastructure such as tap water, electricity, gas, internet access, and roads is now available in their homes. The living quarters have evolved from simple Bengbeng bungalows to spacious and bright buildings or even two- or three-story villas. The previous sheep and cattle shelters have transformed into modern automated feeding

Residential Houses in 1960s

Brick Houses in 1980s

Houses with Tiled Walls in 1990s

Current Spacious and Bright Houses with Courtyards

(Photos source: Wechat Official Account of Ushen County Publication)

farms. The environment of Ushenzhao has significantly improved as herders transitioned from a nomadic lifestyle to settling down and establishing profitable family farms. The scenes of sandstorms burying houses, people and livestock suffering from cold and hunger have disappeared.

I am now 86 years old, and after retiring, I often return to Ushenzhao Town. During the past twenty years since my retirement, I have made eighty trips back there to check on the sand control situation and serve as an advisor to those in need. I also participate in tree planting activities. The trees planted back then are now so big that several people would have to hold hands to encircle them. Now, from the central government to local governments, everyone is very concerned about the desert control. In the past, there were even people who opposed it, but now everyone is taking action. Everyone has this awareness; even children no longer break branches of trees.

Seeing the green trees and abundant of cattle and sheep, not only the herders' lives free from worrying about food and clothing, but they are also becoming increasingly prosperous. My heart is truly happy and relieved because our past efforts were not in vain. I often think that Ushenzhao's pioneering path of desert management and grassland building over half a century ago was absolutely correct. It is a path towards sustainable ecological development. My choice was right; I joined the Party when I was twenty, and for 66 years since then, I have been a Party member without letting down either the Party or the people throughout my life. It has been worth it!

Set Sail, Red Ship

—Memories of the Residential Transformation in My Hometown

by Cao Rui

The phrase "Set Sail, Red Ship" is said about my hometown—Red Ship Town. During the Yongle period of the Ming Dynasty (1368-1644), a large red official ship often docked here, so this place was named "Red Ship Ferry", and gradually it developed into "Red Ship Town".

I was born in Red Ship Town, Juancheng County, Heze City, Shandong Province in 1998. The town is located in the southwest plain of Shandong Province in the east of Juancheng County, bordering Yuncheng County. It is also the hometown of Sun Bin, a renowned ancient military master who was born approximately in 316 BC during the Warring States period. My hometown is situated between Acheng Town and Juancheng County. In the late Yuan Dynasty (1271-1368), families with surnames such as Xue, Guo, Xu, and Yang settled down here and gradually formed this village. Originating from a pier on the shores of the Hu River (renamed "Zhaowang River" during Qing Dynasty from 1616 to 1911), it was initially named Hong Chuan Kou (meaning "Vast River Ferry"). During Guangxu Emperor's reign (1875-1908) in Qing Dynasty, Zhaowang River was transformed into a flatland by the deluge of sludge, and subsequently this area was renamed Red Ship.

In 1957, Red Ship Township was initiated followed by the setting up of Red Ship Commune in 1978 before officially establishing Red Ship Town in December 1983. Since the founding of the People's Republic of China, the habitat environment in Red Ship Town has taken on a new look. What follows below are stories from my grandparents', parents' and my own experiences that describe transformation and rebirth of my hometown throughout three generations.

After the establishment of the P. R. China, with the recovery of the rural economy

Photo of the Old Earth Adobe Houses

and the increasing income of farmers, people began to consider building houses to improve their living conditions. However, they had limited options and mostly followed conventional styles of earth cottages. The living conditions were poor, with three generations residing in two dilapidated adobe houses. Despite being crowded, there wasstill a happy and harmonious atmosphere.

In 1978, the nation began implementing the Reform and Opening Up policy. Specifically,the Household Contract Responsibility System was initiated in rural areas. Since then, every family has devoted all their enthusiasm and energy to cultivating their allocated farmlands, resulting in the resolution of food shortages within just one year. Two or three years later, farmers' income increased significantly, leading to a boom of housing construction. Numerous households started constructing their new "brick and wood structure" houses in the rural areas, making the beginning of Red Ship Town's second phase of housing development.

My father and his two brothers engaged in the livestock business in 1990. With the entire family's hard work, we finally moved into a new red brick house with a courtyard five years later. The earth courtyard walls were transformed into brick walls, while the low and dark adobe cottage was converted into a spacious and bright brick house. The courtyard was essentially a quadrangle style enclosed by surrounding rooms and walls. The whole structure consisted of main rooms, auxiliary rooms, courtyard walls, a gate, a screen wall, a latrine and a dooryard called "Dang Yard" by locals.

My father and his two brothers all got married in this red brick house. Prior to the marriage ceremony of my younger uncle, my eldest uncle and his wife moved to a separate yard, and my grandparents lived with my younger uncle's couple in the main rooms, while my parents resided in an auxiliary room. According to my parents' recollection, at that time, the interior furnishings of our home were simple, consisting of a sofa, wooden cabinets, an old fashioned square table for hosting eight people (known as "Eight Immortals Table" in Chinese), armchairs, small wooden benches, a wall clock, and so on. The walls were plastered with white mortar and the interior floors were paved with concrete. New Year drawings, drama pictures, and family photos adorned the walls. My father was a young stylish man who managed to purchase a bicycle, a sewing machine, a watch and a radio for our family.

My memories of the old courtyard stem from the family reunion holidays when my siblings and I played in the yard while the adults cooked in the kitchen. The tall chimney emitted wisps of white smoke, accompanied by children's laughter, floating up towards the locust tree in the yard, higher and farther.

With the increase in attention and investment in rural development, the living environment of Red Ship Town has undergone dramatic changes. The government has increased investment in infrastructure construction, building wide and flat cement roads with installed street lights so that the village is no longer shrouded in darkness during night. At the same time, the government actively promots rural

My Mother Sitting in Living Room in 1995

My Father Standing at the Gate of "Dang Yard" in 1996

My Mother and Eldest Aunt in the Living Room in 1997

Daily Activities in "Dang Yard" in 2006

housing renovation and encourages farmers to build safe, comfortable and beautiful new residential buildings.

In 1998, in order to promote economic development, Red Ship Town developed land lots on both sides of the provincial highway and constructed residential buildings with street frontage. In 2000, my parents relocated to such a two-story building with three rooms on each floor. At that time, I was just two years old and my earliest memories of life began right there. After the birth of my younger brother in 2004, our family consisting of four members moved to the second floor while utilizing the first floor as a living room and garage.

What I looked forward to most every day was watching TV while having meals on the first floor. Now,when I reminisce about my educational enlightenment, it actually came from cartoons such as *"Journey to the West"* and *"Calabash Brothers"* that were played on the DVD player in this residence. Modern appliances were gradually added to my house: a Simmons bed replaced my parents' dark green earthquake-proof wedding bed;decorative cabinets, a glass coffee table, a complete set of sofas, a refrigerator, a washing machine and other modern items entered my home one by one. Additionally, electric lighting was also installed in the upstairs rooms.

Despite being adjacent to the provincial highway, vehicles were not popular during the early 21 century, resulting in little traffic on the highway in front of those residential buildings. Therefore, my friends and I held bicycle races almost every day, chasing each other excitedly. What a lively scene it was! Behind these street-frontage buildings was a large piece of farmland, and every year when the wheat ripened, the whole world turned into a golden paradise.

I completed my nine-year compulsory education in the vicinity of the home building. Here is the story:at the age of 6, I started my first grade at Red Ship Primary School (later renamed Central Primary School of Red Ship Town), which was situated directly behind my home, across a two-acre field. In today's terms, my home was a type of "school district house". As an independent and adventurous child, I used to ride my bike to and from school every day when I was young. After graduating from primary school, I enrolled in Red Ship Secondary School. Interestingly enough, this school happened to be situated right in front of my home, just across the provincial highway.

Three years later, I went to the county to further my education. At that time, high school studies were quite demanding, and I could only go home once a month. Going back to my beloved home felt akin to reconnecting with an old friend, and it changed from being a daily routine to a monthly eventuality. Each time upon returning from school, I invariably observed certain alterations. Gradually, the provincial highway became wider and wider, with more and more vehicles travelling through it. After being admitted to university in 2017, I could only return home during the winter and summer holidays. From then on, my hometown was associated solely with memories of winter and summer; there were no longer any

A Photo of Myself Standing in Front of the Residential Buildings with Street Frontage in 2002

Younger Brothers and I in the Living Room on the First Floor in 2007

spring or autumn seasons for me. My memories of those street-frontage buildings were stuck in that hot summer when my friends and I played in front of them after taking the college entrance examination.

Since 2009, Shandong Province has vigorously promoted the construction of rural communities, and an increasing number of villages have constructed new communities with unified planning. In 2018, Red Ship Town actively responded to the campaign to accelerate the pace of constructing new rural communities. Our family is also honored to be a witness and beneficiary of this great transformation. In October 2019, during the golden autumn season, my family moved into a spacious and bright community building. The new home has three bedrooms, one living rooms, one dining room and two bathrooms, with an indoor usage area reaching 128 square meters. It is spacious and comfortable. The living room is connected with the dining room, creating an open space with satisfactory permeability that make people feel particularly comfortable staying inside. In addition, there is a 35-square-meter garage, which not only provides convenience for parking private vehicles but also facilitates daily life activities. This improvement signifies more than just a transformation in our living environment; it represents a leap in quality of life and an enhancement in our overall well-being. The moment we moved into our new dwelling was filled with joy and contentment for the entire family as it is not only a place where we live but also a sanctuary resting our soles upon it. We are eagerly looking forward to spending more wonderful time here while creating precious memories in our new abode.

Living Room of My New Home at Red Ship Community

Dining Room of My New Home at Red Ship Community

Red Ship Community features a humanized design. The wide roads are unimpeded,with sufficient parking space, making parking super convenient for residents and visitors. Trees on both sides of the roads provide shade for chirping birds and fragrant flowers, creating a beautiful picturesque environment. Neat buildings with unified designs present a harmonious outlook, and the front row buildings are equipped with street frontage spaces to encourage villagers to run business such as supermarkets and restaurants, adding convenience and vitality to the community.

Different residential buildings have been designed to meet the diverse needs of various families within the community, ranging from 5+1 multi-storey residential buildings (consisting of 5 floors plus 1 attic floor) to high-rise residential buildings with 12 and 17 floors. Each building is equipped with elevators and underground parking spaces, making residents' daily routines more convenient. Gratefully, solar panels have been installed in each household, reducing electricity costs while also contributing to environmental protection efforts. Additionally, significant efforts have been made in infrastructure construction in the Community. The pavements are wide and flat, water and electricity supply is stable and reliable, and the sewage treatment and garbage collection system is perfect. All these efforts ensure the comfort and health of residents. The newly built People's Cultural Square provides a great place for leisure activities and entertainment, while the sports and fitness equipment allows residents to exercise at any time and maintain good health. Today, the Red Ship Community has transformed into an ideal home for living, working, and tourism where we can enjoy modern living facilities, experience a strong sense of community atmosphere, and witness rapid changes taking place in our hometown with each passing day.

In the glorious moment of celebrating the 75th anniversary of P. R. China, the habitat environment of Red Ship Town has also entered a new chapter. Today, Red Ship Town is already a beautiful, livable, ecological and environmentally friendly modern town. In our community,there are wide and flat roads, well-designed

Street Frontage Buildings in Red Ship Community

12-Storey Residential Buildings in Red Ship Community

Party Affairs Service Center in the Red Ship Community

Solar Panels Installed on the Residential Buildings in the Red Ship Community

and elegant houses, lush trees,and blooming flowers that decorate a graceful environment. The community consistently provides efficient services to enhance the happiness of its residents. From being primitive and backward in the early days of regime's establishment to becoming modern and livable at present, the improvement in the habitat environment of Red Ship Town serves as both a reflection of societal development over time and an example showcasing improved living standards. Looking towards the future, Red Ship Town will continue to adhere to green development principles in promoting its habitat environment. I believe that with the joint efforts of all villagers, Red Ship Town will continue its historical journey sailing forward into a new chapter for tomorrow; transforming into a showcase of successful rural revitalization in China.

Under the Sunshine

by Chen Jintai

Everyone yearns for a home, a haven of warmth and care, and a perfect shelter from life's storms. When speaking of housing, it becomes evident that it is an essential necessity for human survival. The evolution of housing has also witnessed the passage of history. "Houses are meant to be live in" — owning a house of one's own remains a cherished dream for many. Similarly, my "New House Dream" also tells a memorable story.

One

Leaving a legacy for their descendants has always been a heartfelt wish of parents, especially the dream of constructing a home dwelling in every Chinese family.

I often heard my father say, "The old house we live in was constructed by your grandpa along with other construction workers, who installed each piece of wood and tile. Your grandpa could remember clearly how many columns and purlins were used." This house was not only a property he left to his descendants but also a testament of his lifetime efforts.

My grandpa was a genuine farmer. In the countryside, if someone's residence was a "straw shack", a shabby "adobe cottage", or a family could not manage to construct a new house, they would definitely be laughed at by villagers and labeled as losers. At that time, the village distributed some wood to each household, providing a good opportunity for the villagers to construct houses. Gradually, wooden house were built up in the village; wealthier families would construct a yard with five rooms. My grandpa was a very proud person. Despite the family being so poor that they could barely feed themselves, he still borrowed some money from relatives and friends. Moreover, through the "job exchange" method popular in local rural areas, he saved on labor costs and eventually did everything possible to build a new house. When the construction was finally completed, my grandpa was so happy that he couldn't sleep for several nights. As the family dined together, he sighed, "We finally live in a new house. Although we owe some debts, as long as we are hardworking and frugal, these debts will be paid off sooner or later."

To repay the family debts, my grandpa left his home village and started a small business outside. He carried a large backpack filled with snacks and sold them on streets as a hawker. He understood that integrity should be the fundamental principle in any long-term business endeavor. Hence, even though he made only a small profit, he acquired some long-term customers. His business gradually became profitable, and some customers were even willing to wait for his goods. My grandpa often told us, "A person is unlikely to become poor from eating too much or wearing nice clothes, but they will become poor from a lack of planning and budgeting." Additionally, my grandma got up early and went to bed late every day, weaving cloth to sell, and her hard work was also rewarded with some income. Thanks to my grandparents' hard work and careful budgeting, they were able to pay off their debts incurred from building the new house within several years.

Two

After the new house was constructed, my grandpa divided the property between my father and his elder brother "equally". However, probably because my uncle's family had a larger population compared to ours, Grandpa did not do really divide it equally. Instead, my family was allocated an obviously "discounted" size with a kitchen section enclosed by a rough fence. There were only three rooms for us, two of which were so cramped that they could barely accommodate a bed each, while the third room was essentially a narrow passageway with a cupboard for grain storage. None of these rooms even had enough space for two people to past each other. In particular, the windows is these rooms appeared quite "unique", as two randomly chosen wooden blocks were installed as window frames and secured by nails, followed by plastic-film seals for insulation and better natural lighting. These windows were intended to serve as an "anti-theft" measure; however, in reality, my home was almost empty with nothing worth stealing. Nevertheless, these wooden blocks provided us with some sense of security.

Despite our family's poverty, my brother and I studied diligently, and we received certificates of recognition from school every semester. However, this also added to the "frustration" of my parents, who did not know what to do with so many certificates. During that time, families liked to display their children's award certificates on the walls of their homes, proudly showcasing their achievements. Unfortunately, our house was so small that there was no space left for new certificates after the walls were fully covered. I came up with a solution by sticking the new certificates on top of the previous ones, completely covering all the walls in our home with these new accolades; thus solving the the problem had been causing distress to my parents. In such a "pocket-sized" home where my brother and I shared one room while my parents lived in another room, our family lived in this way for a long time.

Three

It was not until my maternal grandparents moved into the city that our living conditions got improved.

My mother's brother achieved good academic records and was able to work and live in town, even affording a house there. Consequently, my maternal grandparents left the countryside and lived with my uncle, helping taking care of his children. Considering our poor residential conditions and the convenience of carrying out farm work in a more consolidated manner, they repeatedly advised my parents to move to their country house. Later on, we moved there. Although it was still a wooden structure, there were more rooms available for living. I still remember that this big house could always accommodate relatives' visits; moreover, the living conditions in this house were much better than before. Therefor, at that time I felt somewhat proud and thought that my parents' choice was correct. The windows of this house were also much more exquisite than before: they were crafted into small lattices set in wood, and then nailed with thick plastic films. This kind of window design belonged to those rural families with rather good conditions who could afford extra expenditure.

Although the house was more spacious, both the interior floors and outside grounds were only paved with mud without cement; therefore, we often slipped or stumbled during rainy days, and our shoe soles would become sticky with mud. Since the countryside was infested with rats at that time, the grounds were filled with rat holes created by these pesky creatures. I remember that my wife's family once came to visit us and encountered this situation; after they returned, they exaggeratedly discussed their experience, especially my wife's aunt who enjoyed gossiping and gestured in air to depict a large rat hole.

Talking about the toilet, there was no separate toilet at that time. Basically, the pigpen and the toilet were mixed in use, or an adjacent area to the pigpen was separated as a toilet for slightly more refined families. Similarly, there was no separate bathroom in the countryside; however, the rural people tried their best to solve personal hygiene problems. Nowadays when I recall those efforts, I truly admire their infinite wisdom: as long as there is courage to overcome difficulties, there will be a way out to conquer any challenges.

Four

Each generation has its own aspirations. While Grandpa worked hard to build a house for his descendants, my parents had a different mindset. Although the living conditions of the family also needed improvement (after all, some villagers of my father's age had rebuilt their houses), they would say, "If it weren't for supporting your two brothers' education, our family could have lived in a new house. Nevertheless, as long as you study hard, we are fine with living in the old house." My mother often encouraged us to study hard in this way. In the eyes of my parents, they would be happier if their children could study well and leave the countryside rather than constructing a new house in the village. Rebuilding a house is a huge expenditure for rural residents; if my parents had actually spent the money on building one, it is likely that both my brother and I wouldn't have been able to complete our studies due to poverty. To support our education, my parents did not use their limited savings to rebuild the house. It is this spirit along with their words and actions that supported me and my brother in "jumping" out of the social status of farmers and changing our destinies. It is worth investing money in educating children rather than constructing a house. Consequently, my father's plan of having a new house remained an unrealized dream.

Five

My brother and I were lucky guys. Through hard work, we successfully gained admission to colleges and secured jobs in cities. At the beginning of my career, I resided in a dormitory provided by my workplace. The building was made of brick and wood, and the dormitory room only had enough space for a bed and a desk with minimal remaining space. I remember that my parents encouraged me to save money to purchase an apartment, even promising to match the amount of money I saved if I managed to buy a house. However, being a young man who typically lacked frugality and saving habits, I belonged to the so-called "moonlight" group—referring to people who spend all their income every month without any savings. Consequently, my parents' subsidies turned out to be a bubble.

Noticing that some of my colleagues had bought new apartments, I became envious. Although I had the dream of buying a house, my meager salary was simply out of reach. During blind dates, women would ask me whether I could afford to buy an apartment in the city, and this problem always made me feel embarrassed. At that time, not only could I hardly afford a house, but also had to be extremely careful with my daily expenses. This disadvantage dealt a blow to my heart and resulted in several failed blind dates.

Fortunately, I met my current wife who didn't choose a spouse based on their ability to afford a house. My wife and I rented an old apartment made of brick and concrete, consisting of one living room, two bedrooms, one kitchen and one bathroom. Th apartment was compact in size, roughly 40 square meters; naturally, each room was small. What impressed me the most was that the roof leaked during rainy days, making the floors wet and slippery. Additionally, there were mosses growing on the ground of the public toilet outside. If you weren't careful enough, you could easily slip and fall. During this period of time, my wife was pregnant, and she almost miscarried due to slipping on the wet floor. Thankfully, we rushed to the hospital for a check-up which confirmed that everything was fine. After this accident occurred, I secretly vowed in my heart that I must afford a house of our own someday. A few days later, my wife and I found a safer apartment available for lease in a residential building on a lower floor with one living room, two bedrooms, one kitchen and one bathroom. We were very satisfied with this new place; in order to secure a stable rental price throughout our long leasing period, we even asked our friends who knew the owner to help negotiate the terms. Finally, we rented this apartment for quite some time until we purchased our own new house.

Six

Buying an apartment in the city had always been our dream. After giving birth and even during breastfeeding period, my wife eagerly returned to work at a kindergarten as she desired to purchase a house early. Despite having to breastfeed while working diligently, her salary remained modest. I felt remorseful for not being able to provide her an easier life. Later on, my wife realized that her income could not help much in alleviating our financial burden of purchasing an apartment; therefore, after careful consideration, she decided to start her own business. We led a frugal life and managed to save a sum of money together with my salary; however, this amount proved insufficient considering the huge expenses of purchasing a house.

As an ancient poem goes, "Where hills bend, streams wind, and the pathway seems to end; Past dark willows and flowers in bloom lies another village." The implementation of the national Housing Provident Fund system has granted ordinary working-class families the ability to afford their houses. With the housing provident fund, I regained my confidence in affording a house. The next step was

visiting various sales offices of real estate developers in searching for an ideal house. Eventually, after several rounds of comparisons, we found satisfaction in a new apartment located in Bozhou district, Zunyi City, Guizhou Province.

At that time, elevator buildings had become popular. Speaking of elevator buildings, it was an incredible thing for elderly residents living in rural areas their whole lives and they could hardly image people easily moving up and down dozens of floors by an elevator. Additionally, the buildings adopted a frame structure that was claimed to have a higher level of earthquake resistance. Balconies were installed with tempered glass, and windows were aluminum frame structures. Residential communities were adorned with trees and gardens, and fitness equipment, swimming pools, kindergartens as well as other supporting facilities had become necessary amenities. Environment and supporting facilities had become the main factors for people to consider when buying a house. People used to emphasize a place that "can be lived in"; currently, the emphasis has shifted to "separated functional spaces". After solving the basic living functions, it is more important to improve the quality and comfort of living in a house, as well as children's education and other factors.

Although our apartment was only around eighty square meters with one living room and two bedrooms, I still felt a great sense of achievement and gratitude in my heart when reflecting upon our efforts and struggles, as well as the support provided by the national provident fund policy. Without the national provident fund system, it would be an unrealized dream for people born into rural families to afford a house in urban areas.

I quickly negotiated with the developer's sales staff, made a down payment of over 60,000 yuan, and then all the monthly installments would be paid through the provident fund. Soon after, the housing loan was approved. By establishing a close connection with the housing provident fund, we were able to purchase our new apartment. My wife and I carefully considered how to decorate our apartment and install various household items; after all, this was where we planned to spend our lifetime. Like most urban families, we also installed air conditioners in our home. With just a switch on these air conditioners when it gets too hot or cold outside, we could enjoy comfortable and constant indoor temperature. Whenever I think about my grandpa's struggles in building his home as well as my father's, I feel fortunate that I have caught up with better times and am enjoying benefits from national policies.

Seven

My younger brother found a job in Beijing after graduating from university. He rented a single room when he started working, and later rented a suite with one living room and two bedrooms after getting married. In 2014, his family eventually purchased an apartment in Beijing by paying the down payment with years of hard-earned savings, along with the support of the provident funds. After buying the new house, my family also traveled to Beijing visit them and stay with them. There are complete facilities including central heating at home, as well as public activity venues, public areas for drying clothes and public charging facilities within the community. These supporting amenities make residents live a more comfortable and delightful life.

We finally bid farewell to those "homeless" days. When our parents learned that I had bought a house in the city while my brother also purchased one in Beijing, they were so happy that they almost danced as if their own dream of owning a new

house had come true after all these years.

After the relaxation of the national two-child policy, my family expanded to four members, and our original one living room and two bedroom apartment became insufficient. Therefore, we planned to purchase a 120-square-meter apartment with installed elevators. The new apartment consists of one living room, three bedrooms, one kitchen and two bathrooms (we intentionally chose to have two bathrooms for convenience as most families do nowadays). The bathrooms combine the bathing area and toilet together; we have installed two water heaters and even an intelligent toilet. What caught my attention at first was the spacious balcony in this new apartment. I now cultivate various flowers and plants on it which bring vitality into our home and make my life more enjoyable. Additionally, I have also bought some exercise equipment and placed them on the balcony; these fitness tools are "inseparable" from me after a day's work because they provide added convenience for maintaining regular exercise.

With the improved national policies, the aspiration for better living conditions continues to be the dream of Chinese families. The living environment and facilities, such as community landscaping and fitness equipment, present a stark contrast to housing conditions in the past and have brought about fundamental changes. Moreover, property services have also contributed to owners' convenience, making residents feel secure regarding logistic issues.

Since we bought our new house, I often accompany my children to play and engage in physical exercise in the community square whenever I am free from work or on weekends. Although the conditions of our living environment are not the best, we are happy and grateful, compared with those unforgettable years in the past.

Eight

In the summer of 2010, a fire burned down my maternal grandparents' house where my parents were residing. Fortunately, my parents managed to escape from the fire. Good people often have good luck; no one was hurt. Fire is merciless, but there is love and blessings. The local government departments rushed to the scene immediately after the disaster and distributed relief supplies. Good deeds will be rewarded with more good deeds. The villagers sent rice, oil, chili and other household necessities, and the local government also helped reduce our family's loss. It was such a touching experience that made me understand that kind-hearted people are always rewarded with kindness. Local residents, relatives and friends donated money and goods to help us overcome our difficulties. My parents temporarily stayed at my uncle's house for a while; after we settled things at home, I brought them to live together in the city where I work now. It truly brings happiness for our family to be reunited.

Now the countryside has been transformed into a new image from its previous uncultured conditions, shedding out-of-date labels such as: poverty, poor hygiene, shabby housing, and unpaved roads. In order to address issues related to agriculture, rural environment and rural residents, our country has prioritized the development of agriculture and rural areas. As a result, the countryside now boasts an enviable appearance with its beautiful natural environment, close-knit community connections among neighbors, increased income levels, and improved living standards. Following the completion of the national "Village Roads Project", driving on country roads becomes a pleasant experience amidst serene natural surroundings compared to noisy cities. The successful implementation of the renovation project for rural dilapidated houses is part of the goal of "Beautiful Countryside Construction" campaign. This initiative has significantly enhanced living environment in rural areas. Villas with red tiles and white walls have also been built up; houses nestled around the mountains and rivers are akin to urban residents' desired "paradise", especially when compared to high-rise buildings in cities. People now travel from cities to enjoy weekends in the countryside where

The Mountainous Village of My Hometown Shrouded in Clouds with Farmers' Houses Dotted among Them. When Looking from a Distance, the Scenery Looms and Appears Incredibly Beautiful.

(Photos provided by Bangyou Peng)

they can appreciate birdsong, blooming flowers, majestic mountains and flowing rivers that together form picturesque scenes which captivate their hearts. Rural tourism is gradually gaining popularity while building houses in rural areas has become a dream cherished by many.

Hometown Village Gets a New Look

Nine

We heard that my younger cousin had built a new house in her hometown, and we couldn't wait to pay a visit when she sent us the invitation. The "villa" cost her family several hundred thousand yuan for construction. The entrance yard was paved with cement, providing a "natural" free parking lot that can accommodate several cars, which was an incomparable advantage compared to cities. The house was constructed with a frame structure, and a few columns supported the entire structure firmly. As we got out of the car, we paused outside to admire the elegant facades of the house. The exterior walls had been painted with real-stone coatings, which presented a particularly magnificent appearance. My cousin warmly greeted us and led us into the house for a "home tour": each floor had an independent living room along with bedrooms, bathrooms, and a kitchen; furthermore, the interior decoration exhibited thoughtful design ideas that differed from popular city styles. All of the guests commented on how carefully designed many new rural houses were and how their larger construction areas made them spacious; moreover, their decoration styles kept up with the "new trend" of time by abandoning the old rural "rustic" styles; as a result, rural houses were becoming increasingly outstanding.

Hometown Village Presenting a New Look
(Photo Provided by Renjun Li)

Ten

My younger brother worked in Beijing and did not return to hometown for many years due to various factors, such as taking care of his child. It was not until his family finally returned after years of longing that he was impressed by the completely refreshed scenery at first glance. He could not believe his eyes when he stepped onto the asphalt roads and walked into the entrance of the village. Villagers had all constructed new houses and bid farewell to the era of living in shabby ramshackle houses. Where has our old hometown gone? I remember this villa used to be a dilapidated wooden house, and that two-story white-walled tiled house used to be a small three-room bungalow, etc. The elegant rural houses not only formed a beautiful landscape, but also served as an emblem of the entire rural revitalization process. We visited the resettlement community, where I explained that it was constructed under the government's unified planning; additionally, free housing was provided by the government for families facing financial difficulties who could not afford their own houses. My younger brother exclaimed, "The rural construction is truly making exciting updates with each passing day, and national policies are very considerate." He urged me to take him to where our old house used to stand, so I pointed towards a bridge saying, "Here is where our grandpa's old house used to be before it was expropriated for urban development."

During the process of rural development and construction, some lands and houses were expropriated. The Party committee and government provided several options for residents to meet their distinct requirements. They could choose either to live in houses provided by the government or they could build their own house on land lots allocation by the government under unified planning. After discussion, our family decided to choose the a land lot to build a new house for our parents; after all, we Chinese adhere to the traditional life creed as "Returning to hometown, the root and destination of life." We, their children, cherish deeply the warm embrace of home with our hearts. With support from incentive policies of the government, our dream of having a new house will soon become a reality. The whole family is exceptionally happy about it. What a promising future for us!

Before we returned to cities, my younger brother was reluctant to part with his thoughtful gaze fixed on the houses in the sunshine of our hometown.

The Improved Living Environment of My Hometown

by Shanzeng Ciren

My hometown, Nyingchi City (formerly known as Gongbu), is situated in the southeastern part of Tibet, boasting an average altitude of 2,900 meters. The region is renowned for its abundant forest coverage, including the famous lush forests at Lulang, the picturesque Peach Blossom Gorge, the emerald waters of the Basongtso lake, as well as enchanting Nanyi Gorge. Moreover, due to its sparse population on the Tibetan plateau in general, Nyingchi covers a vast area of 117,000 square kilometers and governs six districts and counties. With an exception to Ba'yi district where the population reached 53,000 by 2015 year data (while other counties have around 30,000 inhabitants), you might perceive these numbers as relatively low; however, apart from Lhasa City itself, Nyingchi already holds the distinction of being Tibet's second-largest city. Within Nyingchi City's Ba'yi District lies both an old area and a new area—while traditional Gongbu ethnic architecture is becoming increasingly rare in rural areas of the old area, while modern urban architecture predominantly characterizes the new area.

Nyingchi receives abundant rainfall and boasts rich forest resources. The minority

The Old Gongbu Streets Viewed from the Sacred Biri Mountain

Nyingchi New Area under Construction

folk houses are constructed based on local conditions using indigenous materials, showcasing distinctive ethic characteristics. The construction of these folk houses highly respects natural surroundings and places great importance on the ancient geomancy theory known as *"Feng Shui"*. Typically, they are built facing the sun, near water sources, and in more elevated terrains. Firstly, lamas are consulted to examine the construction site based on building principles in order to determine a suitable location and perform a consecration ceremony. Secondly, an auspicious day is chosen by the lamas to officially commence construction. Among the local architectural styles found in Nyingchi, Tibetan dwellings and Menba People's dwellings exhibit more prominent regional and cultural characteristics. Nyingchi area is home to multiple ethnic groups with Tibetans being the main community. Apart from Tibetans, there are also Menba, Lhoba, Nu and other ethnic groups residing here. Their lifestyles and religious beliefs retain strong traditional elements while embracing a unique national customs. Ancient legends intertwine unsophisticated folk traditions along with clan totem worship practices and religious myths that add a layer of primitive yet mysterious charm to these ancient ethnic groups and their distant habitats.

Overlooking the Sacred Benri Mountain of Nyingchi

The residential buildings in Nyingchi typically feature herringbone roofs and rectangular layouts, with doors facing east or southeast. These roofs are constructed using long and thin planks without joints, secured with stones to prevent them from being blown away by wind. The local residential buildings mostly consist of two stories and are commonly equipped with stone walls, wooden beams, and wood floors. The ground floor is usually used as a space for livestock or storage for heavy debris, while the upper floor serves as the living room, kitchen, etc. Underneath the herring-shaped roof, fine feed and dried peaches along with other fruits can be stored; alternatively, lightweight debris such as fur can also be stacked.

Due to the low altitude, rainy weather, and high humidity in Metuo County, the traditional houses of the Menba people are usually elevated wooden structure (also known as "Stilted Buildings"). These houses can reach heights of over six or seven meters, with some even reaching over ten meters. In addition to the stone walls,

Granny and Granddaughter Playing in the Courtyard in Nyingchi

A Local Folk House Surrounded by Flowers in Nyingchi

most parts are made of wood and don't require nails for joints, giving them a lofty but solid appearance. These stone-foundation buildings enjoy the advantages of being resistant to earthquake and moisture. Typically consisting of three floors, they have ground floors used for livestock, middle floors for residents' activities such as accommodation and hospitality, and third floors for storing feed, grains, husks and cold wines.

Recently, Nyingchi City has implemented a tourism-driven economic development strategy, attracting visitors from all over the world with its exceptional natural resources and rich cultural heritage. By encouraging farmers and herdsmen to

Original Folk House of Menba People

A Beautiful Lhoba Girl with Her Family of Four Members in Nyingchi

embrace new ways of lifestyle, the government has provided support for them to establish businesses in scenic areas and engage in tourism services such as transportation, catering, and merchandise sales. Simultaneously, significant efforts have been made to enhance infrastructure through comprehensive policies and measures aimed at accelerating information networks construction and promoting industrial growth. This has led to the establishment of the "Smart Tourism · Rural Tourism Information" platform for efficient market operations. With favorable national policies as well as supportive efforts from various provinces and cities across our country, I firmly believe that my hometown will continue to prosper in the near future, bringing wealth and happiness to its people.

From Village to City Square: The Residential Journey of a Barefoot Doctor's Family

by Dong Xuewen and Jin Yanxia

"Our children have all been sent away, and the grandchildren have gone to college. We are familiar with these old places, and there is no better place for us to retire than here." Said Doctor Dong and his wife Master WorkerJin in an interview, both of whom were over 80 years old at that time.

Our Farmhouse on the Yinbei Plain of Northern Ningxia

Before moving to the city, my family resided in Yanzi Dun Township, Huinong County, Ningxia Hui Autonomous Region. The name of Yanzi Dun Township originates from an ancient beacon tower surrounded by earth walls called "Yuanzi Dun" (meaning "Courtyard Piers"). Later, swallows nested on these piers,so it was renamed "Yanzi Dun", which means "Swallow Pier". This is my hometown.

I am a villager from Waixihe Village. I worked as both a village barefoot doctor (translator's note:a kind of uncertified doctors who are mostly peasants providing rural medical services with limited professional training)and a teacher at the township middle school. Villagers always referred to me as Doctor Dong or Teacher Dong. My wife served as a midwife in this village, and everyone called her Master Worker Jin. Our farmhouse was located on an earth terrace named "Zhuangtaizi" in Chinese.

At the end of the 1970s, I was nearly 40 years old. Over the years, I had collected and accumulated various essential materials for building a house, such as large and small stones, wires of varying lengths, bits and pieces of bricks, several beams, and a bunch of rafters with different thicknesses. When the Village Commune building was being demolished and relocated, I hastily bought some dismantled gates and window frames. After contemplating the plan for a long time, I finally felt confident that it was time for me to construct my own house according to my own ideas.

I chose to carry out my plan on a site adjacent to the old terrace and invited strong villagers to help construct an adobe house according to my own design concepts. I treated them with delicious food and drink every day. Upon completing my house, I was able to live independently from my parents and siblings. Glass windows replaced paper-pasted ones, making the new adobe house bright and outstanding. A few years later, on the nation's first Teachers' Day, the local government awarded me a quota to buy a TV set which I added as part of my house property. The towering TV antenna emerged as a prominent feature amidst the row of residences, while laughter from village children could be heard from time to time as they often gathered at my house watching TV. At that time, their favorite episode was *Yuanjia Huo*, a tale about a renowned Chinese martial arts master during the late Qing Dynasty.

The Brick and Tile House of Shizuishan Mining Area

In the late 1980s, as our children grew up, our eldest son started his career at the municipal hospital, and our two daughters also began studying at a nursing school. Due to more frequent commuting between the city and countryside, we experienced much inconvenience regarding transportation. Therefore, we considered moving to the city for the sake of children's career development. Undoubtedly, our dream destination was Shizuishan City, a place renowned for its abundant coal resources in China. However, years of mining production had resulted in many subsidence zones in the city, posing risks for residents. Nevertheless, it remained an attractive place for rural people to live. My wife, eldest son and I decided to use all our savings along with borrowed money to purchase an all-brick house. Luckily, this houses had a large courtyard spacious enough to store some rural debris brought over from the countryside. Carts, shovels, and other items used for feeding pigs were not discarded; insteadly, they were utilized in our urban dwelling for several years. It was in this brick house that my two daughters got married and my youngest son successfully enrolled in college education.

Our courtyard was located adjacent to the back door of No. 1 Middle School in the city. In the late 1990s, it was regarded as a "school district house". Many aspiring high school students chose to leave the crowded school dormitories and resided in nearby residents' homes for an improved living and studying environment. At its peak, over 20 high school students rented our home as their dormitories. We were concerned about these young people's growth during their stay. We lived here until the end of the 20th century and gained some knowledge on managing rental properties.

Apartment in the Central Square of the City

In the early 2000s, as living conditions improved, more coal miner families relocated to Yinchuan, the prosperous capital city of Ningxia Hui Autonomous

Region. Consequently, people living in those subsidence zones were given the opportunity to gradually move into the central district of Shizuishan City; thus our family finally settled in an apartment at the central square of the city. By residing here, we are able to enjoy an urban lifestyle with convenient transportation and shopping facilities while participating in various group activities like local citizens do. Underneath warm sunlight, many elderly people gather together and frequently discuss ways to maintain a healthy body and mentality.

Wenjing Square of Huinong District in the City

After more than 40 years of hard work, we have finally found an ideal place for retirement. For many years, I have cherished the trees in front of our home, and it is so nice to be able to sit peacefully in the square while listening to the bell chiming from the clock tower at the city center. Precise time perception holds significant importance for elderly people. Once my wife mentioned that despite being advanced in age, she could still sense life's vibrancy while watching people joyfully dancing in the square.

Perceptions on the Transformation of My Residential Environment

by Duan Meichun

Born in 1987, I was raised in Shiyan Village, Yufeng Mountain Town, Yubei District, Chongqing City. At that time, the village was called Gulongba Village and belonged to Shiping Township, Jiangbei County. Reflecting upon the first half of my life, I have resided in six different residences apart from school dormitories and work unit accommodations.

During my early childhood, I lived in an old ancestral house and still retained some vague memories of it: earth-rammed walls, a large wooden gate secured with a traditional latch—upon which I had to climb a stool to unlatch. In front of the gate were stone steps numbering around twenty or thirty, leading up to an expansive bamboo forest. As children, we used to play and climb the trees together. I remembered the spacious courtyard that accommodated over ten families of relatives; despite being crowded, it was full of a lively atmosphere!

In the year 1990, my family began constructing a new house, which remains the most memorable dwelling in my recollection. At that time, my father worked as a cook and spent an entire year working in the county town, only returning to countryside during busy farming seasons. On the other hand, my mother stayed at home to take care of me and cultivate our lands.

Since fire bricks were not readily available then, we hired stonemasons to carve stone strips from rocks located at the entrance of the village. Subsequently, we purchased steel bars and other necessary materials before engaging workers for foundation excavation, wall erection, beam placement and tile installation. Eventually, in 1991 we completed a two-story building within one more year's time. The building measured an area of approximately 100 square meters—it was also the first multi-

The Overall Appearance of the Old House in My Hometown and the Carved Lion Statues Positioned in front of It

story building in our local area.

On both the first floor and second floor of this building, there were three rooms each. The grand living room was situated at the entrance on the first floor, serving as a spacious public area for family feasts around a large round table. Adjacent to it on the right side was a stone granary utilized for stacking grains; while on its left was a room for storing firewood and debris, housing an old cabinet from a former landlord's house that had been distributed for free in those revolutionary times. Adjacent to it was a kitchen with an attached pigpen below it, along with a stairway providing access to the second floor.

The Living Room with Photos of My Deceased Grandparents Hanging on the Wall

The Stone Granary in the Warehouse

The Stove and Stairway in the Kitchen

On the second floor, the first room was my bedroom, where there stood a wooden bed gifted by my aunt. The second room was my grandparents' bedroom, furnished with combined cabinets and a sofa bed. The innermost room served asmy parents' bedroom. These three rooms were connected by an in-line balcony adorned with hollow-out cement patterns on its guardrails.

My Bedroom

My Grandparents' Bedroom

In front of the building, there was a cement yard where unhusked rice could be dried in the open air. A pair of stone lions, carved during the Republic of China era, were placed in this yard, which was surrounded by ivy-covered walls. A stone-slate alley to the right of the yard separated our yard from that of the neighbors. Behind the building, there was an open space used as groundwork for our old ancestral house, where I planted some hardy banana trees.

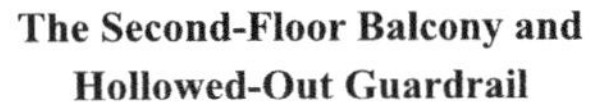

The Second-Floor Balcony and Hollowed-Out Guardrail

Stone-Slate Alley Between the Neighbor's Yards

The Hardy Banana Trees and the Ficus Virens Trees

From the windows on the second floor, I had a view of these banana trees and behind them stood a big ficus virens tree. The ficus virens tree was so massive that it would require two adults to barely encircle its trunk together. On an adjacent open space near the kitchen, we planted flowers along with other plants, such as Cyperus Linn. When I was a child, there used to be a saponaria tree in our backyard where I often picked saponins for hair washing or gathered birds' eggs from nests nestled among its branches. Later on, I planted a new camphor tree there and now it has grown tall and straight.

Looking out of the Window from Second Floor and a View of the Ficus Virens Tree

In 2002, I enrolled in a junior high school located in the county town, while my mother also moved to town for work. As a result, our time spent residing in this house significantly diminished. Hence, we decided to lease it to a construction team engaged in building a nearby highway bridge. During summer vocations, I would occasionally return for about ten days; however, as time went by, our visits became less frequent and the property gradually fell into disuse.

Since moving to the county town, we had been renting and living in a poorly constructed multi-story building owned by locals without any authority approval. This structure was located on land that would later be utilized for constructing a new airport (later known as Jiangbei Airport), where many similar dilapidated buildings were built and primarily leased to peasant workers' families who had relocated from the countryside to town in order to make a living. Due to its bustling yet run-down appearance with numerous peasant workers, locals referred to this area as "Peasant Street".

The residential density in such a structure was exceptionally high. Typically, these buildings consisted of four to five floors and were closely positioned; the gap between each building measured merely 1-2 meters wide, devoid of any trees or greenery. The living rooms were confined, and the kitchens were located in communal hallways furnished with gas stoves. A shared toilet served two neighboring families.

We rented two rooms: the inner one spanned about 10 more square meters serving as a bedroom, allowing for just a double bed (a simple piece of wooden or bamboo board) and a desk, leaving little space for movement. Clothes were packed under the bed in paper boxes, while a small folding table was placed on top of the bed during mealtimes. Through the window, from time to time, I would gaze absentmindedly at pigeons gathering on the rooftops across the street. The outer room was even smaller —more like a wider hallway—barely enough to accommodate a wooden plank bed about one meter wide. At that time, my grandmother and I slept side by side in the inner room while my mother slept in the outer room. My father primarily resided in his work unit's dormitory, occasionally joining us to rest in the cramped outer room where it was even difficult to turn over comfortably. It was truly a tough period when my utmost aspiration was to possess a private space.

After living here for about two years, my father rented another nearby apartment with its size at least tripled. Spanning about 40 square meters, it boasted separate spaces for a lavatory and a kitchen. My father bought a wooden board to build a partition wall in the larger room, so that my grandmother and I could share the rear chamber. Despite the absence of windows, the rear chamber provided my grandmother and me independent single beds and even a designated desk for my study.

This abode served as my residence throughout the high school education lasting three years. Meanwhile, my parents resided in the outer room furnished with a double bed, a desk, and even a large folding table that no longer required folding up. Fond memories resurface from when I hosted thirteen classmates at home during our graduation season in third year; we placed the surface board of our ancestral round table atop the folding table to successfully entertain all guests. During those years, one of my favorite pastimes at home was staying in the lavatory reading *"Dream of Red Mansions"*, as closing its door granted me much-needed solace within an independent private sanctuary.

I successfully passed the college entrance examination and proceeded to pursue my studies in Beijing. Spending a significant portion of my time residing in the dormitory, I slept on the upper bunk with a desk below. Five or six students shared one dormitory, and we got along well with each other, which facilitated a delightful campus life. My journeys back home were limited to summer and winter vacations exclusively.

As graduation approached, my father informed me that he had purchased a second-hand apartment adorned with modest furnishings. Fortunately, the community boasted an appealing environment. Over ten high-rise buildings dispersed in the community at intervals of 50 or 60 meters from one another. The greenery was well-maintained, featuring trees, lawns, sports fields and swimming pools. Garages were all constructed underground. Each evening when I stayed at home, my preferred pastime was leisurely strolling around the community.

The buildings typically consists of ten stories, with each level accommodating four or five apartments and being equipped with elevators. Our abode resided on the fifth floor, a standard two-bedroom suite comprising one living room, one dining hall, one bathroom, one kitchen and a balcony, totaling over 80 square meters in size. In my father's opinion, we directly relocated from a slum to a high-quality community. However, regrettably, a major concern arose due to its proximity to a bustling road.

Moreover, as the road was uphill, buses and trucks refueling generated particularly noisy sounds. Nevertheless, in comparison to the indigent residences in "Peasant Street", these challenges seemed trivial.

In 2016, as part of the demolition campaign for rural development in my hometown, the old house was demolished and now stands as a vacant lot. Only the camphor tree adjacent to the house and the ficus virens tree behind it remain standing upright, just as they were before. Each member of our family received over 200,000 yuan (appr.27,600 USD) as compensation for demolition. We spent this money to purchase a three-bedroom suite in the same community and sold the initial one.

The new suite is situated in a prime location within the central area of the community. My favorite spot is at its entrance where there is an exquisite small pool adorned with a rockery and accompanied by a vibrant young ficus virens tree beside it.

I got married in 2015, and my small family currently resides in a different community within the county town. The dwelling is a two-bedroom suite with a large balcony that covers half of the suite perimeter, providing an expansive area exceeding 30 square meters. This arrangement offers not only a splendid view but also facilitates excellent air circulation. Whenever we have guests, I would gladly give them a tour around my abode.

I am exceptionally satisfied with the current abode. Having originated from countryside with a difficult start, we have struggled all the way through and managed to live a much better life residing in qualified residences. What an achievement!

My only previous concern revolved around the size of our suite, as it might not have been spacious enough for the coming of children into the family. Therefore, I managed to purchase a three-bedroom suite near my workplace. At present, I just take midday naps there on workdays. Once our children reach school age, we can move into this new suite.

Looking back on all these residences, despite the satisfactory residential conditions and community environment we now enjoy, I still hope to have the opportunity to live in that old ancestral house in my hometown. It is not merely about owning a detached property, but rather about fulfilling the dream of embracing a pastoral farming lifestyle:engaging in work at sunrise and finding solace at sunset, while savoring a tranquil, poetic and traditional countryside.

Transition of the Human Settlement Environment of Zhoutaizi Village

by Fan Zhenxi

I am a native of Zhoutaizi Village. The Village belongs to Zhangbaiwan Town of Luanping County. Located to the north side of No.112 Route Highway and Luanhe River,it is 29 kilometers away from the County in the west and 40 kilometers from Chengde downtown in the east. At present, there are 7 residential groups in the village, including 715 households with 2,300 people.

After being demobilized from the army for over a year, I was elected as the Secretary of the Party Branch by all the Party members of the Village. When I got into the saddle, the Village's collective economy had no money, and even owed more than 80,000 yuan (appr.11,072 USD) of debt. Despite our desire to make changes, nothing could be done. In order to change this situation, I reclaimed a village mine that had been operated by my second brother (Mr. Fan, Zhenli) and brought it back into the collective economy. My second brother became very angry and broke off relationship with me. In 1992, I was diagnosed with blood cancer and needed 100,000 yuan (appr.13,840 USD) for surgery. At that time, villagers were still not rich, but every family contributed some money, managing to raise 100,000 yuan (appr.13,840 USD) for my medical treatment. During that life-or-death moment, my second brother even donated his bone marrow to save me. I was touched deeply and silently made a vow: since my life had been saved by the whole village, if I survived this ordeal, I would do everything to benefit all fellow villagers and help them lead better lives.

In 2001, the Village initiated a plan to transform the old village into new residential communities. In those days, the annual net income of Zhoutaizi Village was less than 2 million yuan (appr. 0.28 million USD). However, the Village managed to construct the new residential communities while simultaneously raising funds without receiving any subsidies from higher-level governments or villagers. The construction was accomplished through our own efforts. By 2003, we had built two rows of 13 residential buildings. In 2004, there were 14 new residential buildings, followed by one more in 2005 and another 19 in 2006. By 2008, a total of 408 units across 49 residential buildings had been constructed, and approximately 80% of households had moved into these new buildings. Additionally, we developed a water park spanning 40,000 square meters along with a central square and a park. We also established a new village headquarter, hotels, teaching facilities and a cultural center. Moreover, loop roads around the village have been constructed, as well as main roads and branch roads within the Village. We also invested over 20 million yuan (appr. 2.8 million USD) to implement the Water-Source Heat Pump project, which features advantages such as no pollution, no combustion, no smoke exhaust, and efficient environmental protection. This project enables smokeless heating in winter and cooling in summer for the entire residential communities and office buildings that cover a total area of 11,000 square meters.

The Habitat Conditions of Zhoutaizi Village in the Past

Street View of Zhoutaizi Village Today

In 2009, the Village continued to carry out new residential construction projects. From April to the end of July, the Village Committee visited and mobilized over 180 households living in the old areas to move to new residential communities, leaving the old houses demolished for new planning. The Village provided subsidies according to relevant policies. For example, each house with a property certificate would be compensated 30,000 Yuan (appr. 4,200 USD) ; furthermore, each house and its ground attachments would be evaluated by professional entities to determine a reasonable supplement compensation. By the end of July, our village had basically completed the demolition task. By the end of 2010, 30 residential buildings with a total area of approximately 55,000 square meters had been completed and became new homes for the villagers. In order to encourage villagers to move into new apartments, the Village introduced a series of preferential measures that assured each of more than 600 households owning a set of apartment. Now, villagers live in spacious, bright, and clean "well-off new homes" just like the city fellows. Supporting facilities including hospitals and schools will also be put into use upon the completion of these new residential communities. Although Zhoutaizi Village is located in a mountainous area, it has taken the lead in several aspects; what is now advocated by the Central Government is what we initiated more than ten years ago, such as renovating residential communities and improving rural environment.

In the same year of 2009, we constructed 100 apartments specifically for the elderly in our village. Each apartment has a size of 68 square meters and is designed to improve the living conditions of villagers aged over 70 years old. Currently, about 120 elderly residents from Zhoutaizi Village reside in these apartments. Elderly villagers above the age of 70 who possess residential registration with Zhoutaizi Village can choose to live in these apartment free of charge. The Elderly Apartment Building is fully furnished with all necessary appliances and furniture. Moreover, the Village provides a monthly pension of 500 yuan (appr. 70 USD) to individuals over 70 years old and 300 yuan (appr. 42 USD) to those above 60, which is a unique benefit offered by Zhoutaizi Village. Every day, staff members monitor the health conditions of the elderly, ensuring timely diagnosis and appropriate measures are taken if any illness is detected. Some rooms have two beds while some have only one. Senior villagers can choose to live with others, or couples can live in separate apartments. Each apartment provides convenient amenities such as independent bathroom and kitchen. Elderly villagers find great satisfaction in staying here, as it relieves them from the worries of aging challenges. They all say that living here leads to a long and happy life. Presently, our village already has two centenarians.

Apartment Building for the Elderly, Zhoutaizi Village

Inside the Apartment for the Elderly

In the east of the Village, we built 100 units of welfare apartments with a size of 80 square meters each, mainly to solve the problem of three generations cohabitation. Before the old village was demolished, big families consisting of three generations always lived together in large houses with three or four suites with separate entrances for older and younger generations respectively, aiming to avoid inconvenience caused by different living habits. After moving to new residential communities, conflicts or inconvenience become inevitable. Therefore, after the demolition and reconstruction of the Village in 2009, a Welfare Apartment Building was constructed especially for middle-aged and elderly villagers aged 50 to 69 years old to have their own independent space even though they already have their own family apartments to share with their children. Thus, three generations of a family can each enjoy their own home spaces. Qualified villagers have the right to live in the welfare apartments free of charge; however, they do not have property ownership. At the age of 70, they can choose to move to the Elderly Apartment Building.

We also constructed a Science & Technology Building with a total area of 6,000 square meters. It serves to accommodate professionals and staff members along with scientific and technical training and exhibition functions.

With the improvement of the living environment, children's education has become a top priority in the Village. A kindergarten and a primary school have been established with advanced teaching equipment and facilities. Teachers are uniformly dispatched by the local government. The kindergarten is as wonderful as those in cities: kids are sent in the morning and picked up in the evening, and they have good lunches and take noon naps at school. We also have a fully scheduled primary school where students receive education before entering middle school. In fact, many other villages struggle to provide this opportunity as their children have to transfer to remote schools for higher grades study of primary education. Now the primary school and kindergarten in our village also accept children from other villages.

Village Kindergarten

Village Primary School

We constructed a Cultural Center in 2007, where villagers can enjoy movies and cultural shows. For example, performances are held during every New Year Festival with accommodation capacity for up to 500 people. The Village has established an art troupe that carries out continuous activities, especially during festivals when activities reach their peak. With increased income, villagers naturally pay more attention to their cultural life. A few years ago, we even hosted the Fitness and Yangko Dance Contest of Hebei Province. We have also constructed a central square and a water park for recreation, transforming the entire village into an environmental park with modern buildings and urban facilities. All these efforts make people believe that our village is no longer like the old countryside.

Village Center

Science & Technology Center

Water Park

No matter what kind of construction is planned in the village, a Villager Representative Meeting will be launched to solicit opinions from villagers. For example, we have prepared four optional plans for the construction of residential buildings; we would adjust the plans according to villagers' opinions; and finally, every villager will vote to make the decision. Even though such efforts may not achieve 100% satisfaction, we have tried our best to obtain as much consent as possible. In summary, Zhoutaizi Village could hardly achieve today's success if without the guidance of government policy, attention from senior leadership, and trust and support from its residents.

Auspicious Clouds Embracing the Blessed Homes

—Changes of Newlywed Dwellings over the Past 30 Years

by Guo Lianjun

I was born and raised in Harbin, a city renowned for its frigid winters and musical traditions in the far north of China. Since my early years, I have harbored a fondness for arts. In the third grade of primary school, I successfully joined the Culture & Art Performance Team organized by the district authority and played the pipa in a Chinese orchestra. Later on, I became a member of the city's art troupe and actively participated in performances, including tour shows in various districts and counties. Upon graduating from university, I secured a position in a state-owned military factory where I became a leading figure in art shows. During this period, our factory's art performance group also assumed responsibility for hosting wedding ceremonies for young workers. These weddings were not only meant to be lively and joyful, but also symbolized commendations from the senior management towards young people. From my early years as an employee until retirement spanning three decades from late 1970s to early 2010s, people's life standards have remarkably improved since China's reform and opening up. Throughout these years, I have been honored to take part in numerous wedding ceremonies and visit the newlywed families. In some way, I could consider myself as an eyewitness to the transformative changes brought forth by this epoch.

In the Late 1970s

The factory at that time would solve the housing issues for the young couples with legal marriage registrations, providing them with assigned apartments now known as "tube buildings". These dwellings were what we bachelors adored and longed for as they truly represented a high quality of living conditions in those days. Each apartment was typically a single room, measuring about 10 or more square meters, which had to be divided into functional areas serving as kitchen and bedroom respectively. However, it was impossible to further divide the tiny space into a dedicated living room or dining room; instead, it had to serve multiple functions. When hosting guests, folding chairs and tables would be set up for meals, and we fondly referred to such a table as "stand aside". At that time, buildings did not have central heating systems: bungalows usually featured traditional brick beds called "*kang*" (translator's note: a traditional brick bed that can be heated in North China); while multi-storied buildings typically had heating walls installed in their compact rooms.

If a man wanted to get married, he had to provide a set of furniture, which included a double bed, a wardrobe, a tea table, a chest of drawers, and a combined high and low cabinet. Anyway, a complete set of furniture should meet the standards of having "36 angles" and "72 legs" fashioned in those days. Normally, furniture was handcrafted by specialized woodworkers. Therefore, woodworkers in the factory were always popular buddies among young people. Sometimes, senior members of a family who were craftsmen would make the furniture themselves and then invited skilled machinists to solder patterns onto the wood surfaces and apply glossy oil finishes. At that time, lime water or plaster powder were used for decoration: rooms and walls were painted white with green or red as decorative accents on the lower part of the walls. That was exactly how a brand new wedding house was successfully prepared.

Combined High and Low Cabinet and TV Set Popular in Early 1980s

Women also had to prepare decent dowries, among which the "three essential items" (later known as "three revolving and one audible") most expected included a watch, a sewing machine, a bicycle and a radio. Prestigious brands encompassed the "Shanghai" watch, the "Bee" sewing machine, the "Flying Pigeon" or "Forever" bicycle, and the "Panda" radio. With such an opulent dowry in possession, a bride truly shone in front of her friends and relatives.

In the 1980s

In the 1980s, the factory began constructing central heating buildings, with each apartment equipped with a flush toilet and a kitchen featuring a fume pipe. The apartments also had spaces dedicated to specific functions such as living room and bedroom. Typically, three-bedroom units spanned an area of 40-60 square meters, while two-bedroom units measured around 30 square meters. Although by today's standards the apartments may seem petite, they were highly coveted by the young people in the factory during that era. Primarily allocated to managers and model workers, these novel dwellings later became home to many employees' offspring who joined the workforce themselves. As they reached marriageable age, I had the pleasure of attending numerous wedding ceremonies in these newly built apartments.

The wedding houses were adorned with exquisite elegance then, featuring wood-grained paper veneer on the suspended ceilings, soft sponge wraps in silky cloth on the upper part of the walls, plaster lines or carved flowers on the wall copings, and synthetic boards on the lower part of the walls. Some families would choose carpets laid on timber floors, while the majority preferred plastic flooring. Furniture preparation had shifted from handcrafted ones by carpenters to ready-made furniture combinations from factories. Families with good economic conditions could afford leather sofas, whereas ordinary households favored sofa beds made of knitted fabric.

Trendy Combination Cabinets and Color TV Sets in the Late 1980s

When men's wedding houses being upgraded, women's dowries were also upgraded to the "new three items": television, washing machine and refrigerator. Even though the refrigerators had only one door, the TVs were black and white, and the washing machines had only one cylinder, the daily life remained vibrant. In the early 1980s, anyone who possessed a television set would attract neighbors from all the streets to gather in their homes. Especially during volleyball matches or broadcasts of Hong Kong dramas, there would be a sea of people. By the end of the 1980s, well-off households could afford 21-inch color televisions that cost more than 3,000 yuan (appr. 415 USD) at the time-while my monthly salary was less than 300 yuan (appr. 41.50 USD).

In the 1990s

In my recollection, many bungalows were demolished and reconstructed during this decade. There used to be many young people in the factory who did not benefit from the apartment distribution policy, but their family bungalows caught up with the era of urban construction. Therefore, they moved from bungalows to multi-storied residential buildings featuring spacious living areas and comfortable conditions. It used to be a privilege for managers and model workers to reside in suites with three bedrooms and one living room, while ordinary households could also make it then.

Busy in a Wedding Ceremony of the 1990s

As the economy improved, young people were able to spend more money. Take the decoration for instance, living rooms usually featured European-style ceilings with crystal chandeliers, and Roman column structures replaced previous plaster flowers. Bedrooms were decorated with wallpaper, while kitchens were installed with stone countertop along with cabinets made of decorative boards. Bathrooms exhibited a sense of vogue through fashionable mosaic paving.

At that time, numerous new-home appliance stores and furniture stores opened, offering a wide range of grades and prices for instant purchase of wedding supplies. Panel furniture combined was in vogue, while solid wood furniture gained renown for its solidity and stability. Solid wood flooring prevailed during this period, with craftsmen matching parquets on-site, and applying oil paint and polish. Everyone desired to replicate the "decent scenes" seen on TV, in grand restaurants, and decent hotels back home, and the overall style exhibited a distinct European influence. The "three items" required of marriage were upgraded to include a color TV, a motorcycle and a combination stereo.

Living Room in a Newlywed Dwelling in 1990s

New Millennium

Since the year 2000, young couples would have been acquiring their own commercial residential properties for marriage. Residential buildings have become taller and more spacious in this decade. Almost every family had only one child, and parents from both sides naturally wanted to spend money on the young couples. During that time, typically the grooms' families provided the dwellings while the brides' families furnished them.

A New Couple of the New Millennium, a Groom Fetching His Bride By Car, the Facade of the Residential Buildings

Decoration was no longer just about casually looking for a team of craftsmen; instead, new couples preferred to first design the spaces by themselves or consult with designers to ensure their desired styles before contracting with decoration companies. The range of decoration styles became more diverse during this period, expanding beyond European style to include traditional Chinese style and even variations within European styles.

An array of home appliances arrived, ranging from plasma TV sets and updated combination stereos as home theaters to famous brands of refrigerators including "Haier" and "Siemens", along with advanced washing machines using drum technology, water heaters, dishwashers, microwave ovens, and electromagnetic cooktops.

During this decade, I transitioned from being a bustling supporter running errands for the wedding ceremonies to becoming an esteemed marriage witness at grand feasts. To be honest, I felt a tinge of envy towards these young people who could lead such a good life at such an early stage. At times, I also pondered over their future responsibilities—how could the young couples manage taking care of four elderly family members? That would pose as a true challenge for the young generation.

During 2010s to 2020s

I retired in 2012. If counted on the retirement time node, I could be regarded as a post-10. Since retiring, I no longer bustled around attending ceremonies of newlywed young couples at the factory. Instead, I found myself frequently consulted by old friends seeking advice on purchasing houses for their children. What impressed me most was that the younger generation had a preference to relocating their residences to the north bank of the Songhua River.

A Bride in Her Parents' Home in the South Bank and Her New Home Community in the North Bank

As people may know, Harbin is a beautiful city that stretches along the Songhua River. Prior to the establishment of the current regime, Harbin residents used to live on the south bank of the Songhua River, while the north bank served primarily as a hub for shipping or flood discharge purposes. However, with the advent of a new era, the city expanded from the southern shores towards southeast. Later, Songbei New Area was officially established in the north bank area with numerous real estate properties offering low prices but maintaining good quality and exquisite environment. However, few native Harbin residents were willing to settle there. In 2010, the Harbin Municipal People's Government officially relocated from the south bank to Songbei New Area, which had prompted many of my old friends to settle down there and take care of their children who had secured jobs in Songbei New Area. It was indeed convenient for young people; however, my old friends had to commune between both sides—their children's home in "north bank" and their own dwellings in "south bank", thus becoming "commuters" after retirement!

I have devoted half of my lifetime for the marriages of other families' children, while my own children live far away from me; occasionally, my heart loses its balance for a while. However, I understand the next generation will surely explore their own way of life as they grow up.

I sincerely wish that the young generation of our country can lead wonderful lives by happily getting married and successfully starting careers!

Home and Away

by He Tingmei

Born in 1989 in Yongzhou City, Hunan Province, Ms. Tingmei He currently resides and works in Guangdong Province. (Interview conducted in March 2024.)

Habitat during Childhood

I was born in Yinjiadong Village, Renhe Town, Ningyuan County, Yongzhou City, Hunan Province. My parents always worked and lived outside of our hometown and rarely returned to stay. When I was young, my parents took me with them wherever they went for work so that I had the opportunity to visit many places. Before starting school at the age of seven years old, I occasionally accompanied my parents on their trips to Guangdong Province, Jiangxi Province, and other places. When it was time for primary school, they sent me back to live with my grandma in Hejia Village, Qingshuiqiao Town of Ningyuan County for three years. During holidays, my grandma would allow me to visit my uncle's and aunt's houses; each holiday meant a home visit to different relatives who also lived in different towns within Ningyuan County but not too far away from my grandma's home; we could take a bus for the visits.

My grandma's house gave me an impression of a small cottage made of mud bricks without a yard. It was located next to the village ancestral hall, where villagers would gather and pound sticky rice cakes to celebrate festivities with a lively atmosphere. During that time, rice was commonly grown in rural areas of Hunan Province. Farmers had the option to plant early-or late-season rice. Some villagers who planted early-season rice chose not to plant late-season rice but to grow tobacco, resulting in each household preparing a tobacco-baking house. These tall structures allowed for hanging tobacco leaves on sticks for preparing flue-cured tobaccos. Since my grandma's kitchen suffered from leaking problems, we used the tobacco-baking house as an alternative cooking space. My grandma and I endured hardships together while taking care of our pigs that were raised in the leaky kitchen. Every morning I would assist with farm work by cutting grass for the pigs. We typically only consumed meat once a week; otherwise, we would buy a piece of lardo and fry it into crispy oil dregs—just adding a little into vegetables during cooking would present hearty meals.

My parents took my younger sister with them while they were working in Guangdong Province and other places. In 1998, my father returned to our hometown and built a one-story bungalow at Yinjiadong Village, Renhe Town, Ningyuan County, Yongzhou City. By the time I was 9 years old in the second semester of Grade Three, my grandma was getting too old to take care of me alone. Therefore, when the construction of the bungalow was completed, my parents decided to take me to Guangdong Province and lived with them.

The Old Mud-Brick House in Yinjiadong Village, My Hometown

Our family rented a house in Shangba Village, Shiliting Town, Zhenjiang District, Shaoguan City, Guangdong Province. Both of my parents worked nearby. Mom worked as a canteen workerat a vocational high school, while Dad ran a waste collection site. The house we rented was quite old; it was made of bricks with outer walls painted with a thick layer of lime that easily dropped dust when touched. It was one of several tile-roofed houses enclosed within a courtyard. The one we lived in was located next to the village ancestral hall where many people could be seen burning incense to worship during traditional festivals.

My parents wanted me to attend a primary school near our home in Shangba Village, but unfortunately it was so overcrowded that students without local residential certificates were not allowed admission. As a result,we had to find another school quite far from our home. The school was located in Xihe Village, Wujiang District, Shaoguan City, which would take me half an hour by bike. Moreover, we had to pay the school 900 yuan per semester as a "Construction Fee" since we were not locals, which was quite expensive for us at that time. During this period of my life, my mother worked while my father took on the responsibility of cooking for me and my younger sister before taking us to school. Sometimes he would complain about how troublesome these childcare tasks could be. I felt sorry about it and decided to start riding to school by myself instead. Being a bold child, one day while my mother was taking a nap, I secretly took her bicycle outside and learned how to ride it. Later on, I asked my father if he could find an old bicycle while collecting scrap materials around town and fix it up for me. Fortunately enough,in the second half of Grade Four, I finally obtained a repaired old bicycle and began riding it to school every day. Since my younger sister was two grades lower than mine, my father continued dropping her off at school and picking her up everyday just like before. That's how we lived in Shangba Village, Shaoguan City, Guangdong Province from Grade Three of primary school until the end of my junior middle school. We seldom returned home to Hunan Province during those years.

Working Outside

In 2005, when I graduated from middle school, I wanted to start working and earning money on my own. Since then, I had traveled to various places in Guangdong Province for work. My first job was as a waitress at a restaurant in Shiliting Town, Shaoguan City, when I was still under 16 years old. However, due to some customers trying to persuade me to drink alcohol, my mother advised me not to continue this job; therefore, I only worked there for less than a month. Later on, through an introduction from an aunt, I worked at a toy factory in Dongguan City, Guangdong Province for half a year. My responsibility was manually sealing the plush toys. The factory provided free food and accommodation, but required me to start working at 8:00 am and finish around 9:00 or 10:00 pm if overtime was necessary. During this period, I lived in a dormitory shared by eight girls with relatively good living conditions including a communal bathroom. However, overcrowding became an issue; for instance, I had to fetch hot water and queue for a small independent space to take a bath. At first, I thought the work was pretty novel but later on found it very boring. The lack of personal time in the factory made me realize that it was actually a hard work. Furthermore, without any opportunities for personal development or learning opportunities, I concluded that I would never work in such a factory again and decided to change my job.

In 2006, my mother contacted an old acquaintance who said she could introduce me to work in Foshan City, Guangdong Province.Since I had grown up in Guangdong and spoke fluent Cantonese, finding a job would be easy for me. She took me to Foshan City, where we lived with her family in a rented house located in an "urban village" on Pingzhou Street, Nanhai District, Foshan City. The house was quite simple and situated on the top floor, with an additional makeshift shed on the rooftop. My aunt had a son who was slightly older than me and occasionally came home. She wanted to arrange a match between us, so she kept postponing introducing me to work and kept me at home. Feeling quite anxious,I went out to look for a job on my own while she was out at work.Eventually, I landed a job at a fruit supermarket and started as an ordinary employee before working my way up to become a cashier.

After securing my job at the supermarket, I decided to move out of my aunt's home. I firmly told her that I had found a job and would be living in the supermarket dormitory. However, deep down, I still hesitated because the salary there was not particularly high, only about 800 yuan per month, although it included free accommodation and meals. The dormitory conditions for supermarket employees were quite good, which was located in a residential building with just a five-minute walk away. It was an apartment with three bedrooms, one living room, one bathroom and one balcony; much better than the factory dormitories. At that time,four girls shared one room, while two bosses each had their own rooms. While working there, I also met my future husband, who was then a manager of the fruit supermarket. Originally catering mainly to workers from nearby factories who would come to buy fruits and snacks after work or on holidays, business was thriving until many factories closed down due to the financial crisis in late 2008. Our boss chose to sell the supermarket away. Meanwhile, I became pregnant but faced opposition from my parents regarding our marriage plans which made me hesitate even more about our relationship. Consequently, I returned to Shaoguan

City, Guangdong Province to give birth to my daughter while my boyfriend went back to his hometown in Hengyang City, Hunan Province.

Later, we got married.In 2009, when our daughter was only six months old, a classmate of my husband's who was working in Shenzhen City, Guangdong Province, introduced him to work there. So we took our daughter to the Guangming District of Shenzhen City to start our business. We took over a canteen at an electronic factory with a workforce of 500-700 workers. We lived and ate in the factory. Since we had no experience in this business, we did not secure an agreement with the factory requiring workers to dine in our canteen. Consequently, some workers chose not to eat there and within just a few months, we failed to continue operations.

In the second half of 2009, we moved to Wanhe Wholesale Market, which was located in Haifeng County, Shanwei City, Guangdong Province (Editor's note: Wanhe Wholesale Market is now known as Wanhe Zhaoxiang Agricultural Products Market), to start our wholesale business in dry ingredients and spices. We stayed in Haifeng County for eight years and experienced diversified living facilities with increasing rental prices year by year. Initially, the entire market consisted of tile-roofed buildings, and we rented three connected shops, each with an area of about 40 square meters. We lived inside these shops directly. One of the three shops was fully stocked with goods,while the other two had partitions; the front halves were used for storage purposes, and the back halves were used for living. One partitioned room served as a kitchen and bathroom, while the other one was used as a bedroom. The shops had roll-up doors installed, and one thing I remember very clearly is Typhoon Usagi in 2013 that tore off both the roll-up doors and the entire roof of the market's canopy—it was very serious. During that period, I gave birth to my son in 2010, and when my children reached school age, I sent them back to my husband's hometown in Hengyang City, Hunan Province, under the care of my husband's sisters. Although I couldn't stay with them all the time due to work commitments, I often took the high-speed trains to Hengyang City to visit them. Before the availability of high-speed trains, I used to drive there myself every time, which would take ten hours.

During these eight years, the market underwent numerous changes. On one hand, damages caused by typhoons had to be repaired, and some old houses needed renovation. On the other hand, there was a need for expansion in the market. Prior to this expansion, there was only a single row of houses in the market. Afterwards, with more shops and tenants, more people came to do business. People could be seen strolling through the alleys of shops in the market. We also relocated to newly constructed shops closer to the market entrance, albeit at higher rental prices. The two storefronts we rented were about the same size as before, and we also leased a warehouse across the street. The new storefronts were constructed of iron, allowing us to add structures according to our needs. As a result, we built double-story structures with attics inside them. At first, we resided in one of the iron buildings for a while but soon found it too hot during summer months. So we didn't stay there for long and rented apartments in a nearby village which was only 5-6 minutes away on foot. We initially rented an apartment on the fourth floor that had three bedrooms and two bathrooms, but found it tired to climb up stairs. Therefore, I decided to move into another apartment on the first floor within the same area that had two bedrooms and two bathrooms.

Wanhe Wholesale Market

B & B Hotel

In 2017, my husband's elder brother, who was working as a scientific researcher in another city, came to Chenzhou City, Hunan Province, on a business trip. He found the natural environment of Dongjiang Lake (editor's note: located in Zixing

Town, Chenzhou City, Hunan Province) impressed him greatly and he believed it had potential for tourism development. He invited my husband to investigate business opportunities together. Considering that our children were studying in Hunan Province but lacked proper parental care, he suggested that we should return to Hunan Province for the sake of our children instead of constantly being away and focus more on their education. Initially, I disagreed with this suggestion; however,when my husband had an accident and required care as well, it seemed there was no other option for us. So, my husband and I decided to relocate to Liangshuwan area of Shoufu Road, Dongjiang Street District, Zixing Town, Chenzhou City and purchased a house there to start operating a B&B hotel.

We purchased a half-finished house with three-and-a-half-story, and decorated it in various styles. Our homestay boasted 10 rooms and could accommodate more than 20 guests, while we resided in the attic. The surrounding environment was pleasantly tranquil, making it an ideal spot for a relaxing holiday stay. We completed the renovation of our B&B in April 2018, conducted trial operations in August, and officially opened during the National Day holiday, which was perfectly timed for peak tourism season. I enlisted some acquaintances to assist with the hotel design and planted numerous roses in the yard; when they bloomed, the entire wall looked stunningly beautiful. Managing a homestay might seem effortless at first glance but actually demanded considerable effort—staying up late being one of its major challenges. Sometimes guests would seek my advice late into the night; to ensure that no orders were missed, I had to respond to questions or requests from guests.

The Half-Finished House in Need of Renovation

The Courtyard of Our B &B Hotel

Life without a Break

However, the market gap between peak season and off-season is too significant, result in unstable income. The homestay's peak season lasts only from May to June every year, extending until the National Day holidays. After that, there are hardly any tourists, and even after the Corona Virus Disease 2019, guests are scarce. Therefore, while running my homestay business, I had been actively sought out other opportunities for family income and personal development. For instance, in 2019 I successfully interviewed for a position at the branch office of Junlebao Company (editor's note: Junlebao Dairy Group) in Zixing City where I worked on business promotion; I also learned how to bake by studying and working at a pastry shop; additionally, I worked as both a receptionist and an instructor at a gym.

By the end of 2023, my daughter had grown older while my husband's health situation had declined, so I had to start working outside again. I took up a position as a greeter and waitress at a hotel in Changsha, the capital city of Hunan Province. Meanwhile, my husband helped manage the homestay in Chenzhou city while also taking care of our two children. On the second day of the Spring Festival this year (2024), I accompanied a tour group as their driver. On the sixth day of the New Year Festival, a friend invited me to go to Shenzhen City with her. Soon after arriving there, I found employment as a hotel foreman. This job also provides free food and dormitory accommodation.

About "Home"

Over the years, I have lived in various places. Among them, the housing experience in Shanwei City of Guangdong Province has left the deepest impression on me. From living in tile-roofed houses to tin houses and residential buildings, I witnessed a significant transformation of the local living environment. However, what remains most memorable are the years I spent living in my grandma's cottage in Hunan Province. It always brings back childhood memories of carefree days spent happily rolling around even in desolate fields. One thing that stands out is my mother returned to visit us and quietly left while I was asleep. Although I spent most of my childhood in Guangdong Province and prefer it for its food, lifestyle, and weather conditions, I still haven't made any specific plans about where to go in the future. Despite having returned to Hunan Province for six or seven years, my heart still yearns for something beyond.

If you ask me why I have multiple jobs and engage in various occupations, I can only say that circumstances leave me with no choice but to seek opportunities and make a living. Currently, I don't have any long-term plans; my main focus is on performing well in my job to ensure a stable monthly income. A few years ago, I faced financial difficulties and incurred debts when purchasing the homestay property. My current goal is simply to repay those debts, and acquire a home of my own, provide a place for my daughter and myself to live. Sometimes during conversations with friends, they ask where my home is and I jokingly reply that it's all over the world since none of the places I've lived so far has been particularly satisfying. The kind of house I want to live in the future, is something I haven't seriously considered yet; there are no specific requirements except for general aspects like convenient transportation, nearby dining options, comfortable living conditions, and easy access to shopping facilities. Ultimately, as long as there's a place called home where we can nestle together peacefully.

Living Housing Experience

by José Manuel Ruiz Guerrero

Torre del Mar

I was born and grown up at early 1980's in a tourist town at the South coast of Spain named Torre del Mar, Málaga province. The literally translation of the city's name into English is "Tower of the Sea", in reference to an old castle which used to surveillance the coast and protect the main city from looting and invasions.
My neighborhood is south-west of the town, 50m away from the beach and it is integrated by three different types of buildings:
a) 2-storey traditional fisherman lime dwellings renovated;
b) 5-storey residential dwelling, built between 1977-1979; and
c) 10/12-storey building for tourist apartment, built in 1975.
My immediate family and I live in the first floor of one of those 5-storey concrete

building, whereas my grandmother used to live in 1-storey fisherman houses, just 30 meters from us. The main façade of our building is cement in material and painted in brown. It is oriented to south-west to a 15m wide street,whereas opposite façade facenorth-east to the fisherman neighborhood. The building would receive much sunlight if it were not because there is a 11-storey tourist apartment just in front; as a consequence, our house is a bit dark.

We are a five members family (father, mother, brother, sister and me) and the 100 square meters area it was enough for the whole family to live comfortably. At the beginning, the house was divided into 3 bedrooms, 1 main living room, 1 secondary livingroom, 2 toilets and the kitchen. At that time, It was a tradition between families to use the biggest area of the house as the main living room for special events like birthdays, meetings or family reunions, and to use a minor space as a secondary living room for the daily activities such as watching tv, reading or sleeping on the sofa.

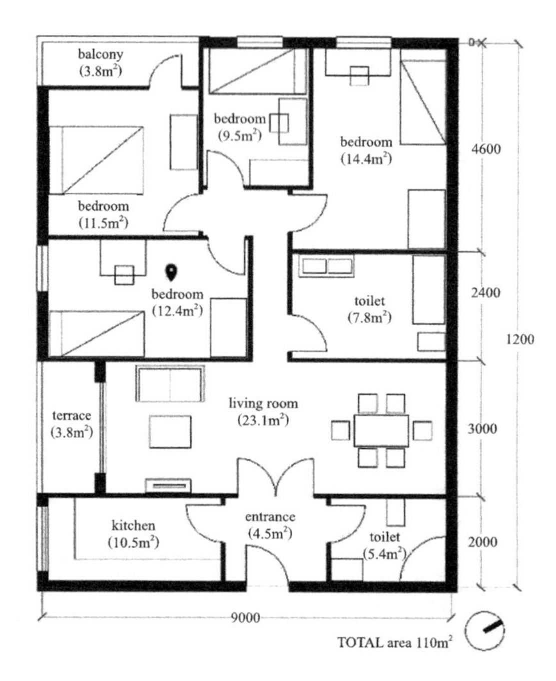

This neighborhood in the 1950's used to be just occupied by a group of 1-storey fisherman white-lime dwellings, one next to each other, forming a block of 200m length and 9m width, with one façade oriented to the sea, south-east, and the

opposite façade facing to the vegetable and potato garden, north-west.
During my childhood, I used to spend my spare time playing with my friends in the street. We had many different areas to play within a short distance: pedestrian square, sport facilities, promenade and the beach were displayed in a distance of just 100-200 meters.

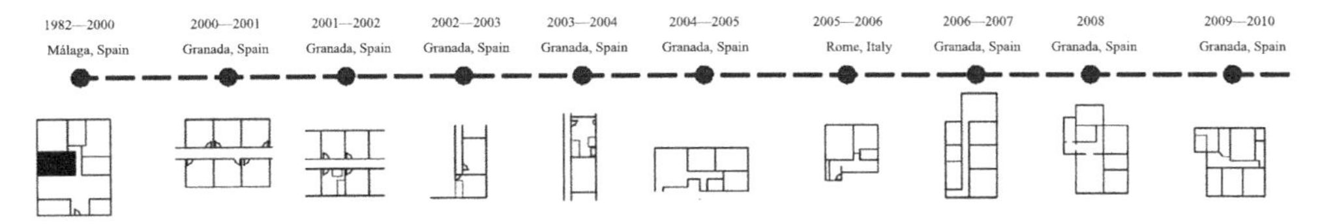

Granada (dominicos)

I had the first big change environment when I went to study Architecture at University of Granada where I had the chance to live four years in a very special dormitory. The dormitory is a 500 years old historic building in the city center of Granada and It occupies an area about 3500m^2, with a square renaissance cloister in the center. It was a luxury to live in this dormitory, not only because every space was generous in dimensions, but also because the variety of functions. In the first floor are public functions areas such as hall, entertainment room, canteen, gym and cloister. In the second floor there is a library, an old conference hall and the TV room. The third floor is occupied by director office. And the fourth and fifth floor are the individual rooms to host the students. Every corridor that leads to those room has a name, and most of these names refers to different city's street names.

According to how many years you were living in this dormitory (your ranked), your room could be $13m^2$, $15m^2$ or $18m^2$. In my first year I used to live in a $14m^2$ room, with one small lavatory and one window faced to south. Everyday around 12, the amount of sunshine it was such, that I decided to set the bed just under the window. Accustomed to living in a dark room in my hometown, I appreciate this sunshine that offers the new dormitory as if they were gold. The room was big enough to have a bed, drawing table for student architect and book shelves. Every wardrobe was built-in, and the wall was white plaster. Every room thermally was very comfortable, warm in the winter and fresh in the summer, and the reason is because the wall thickness, about 50-60cm. The toilet was common toilet out at the end of the corridor, shared by 12 students. Every space mentioned such as rooms, toilets and common areas were very clean because we used to have staff in charge of taking care of the cleaning.

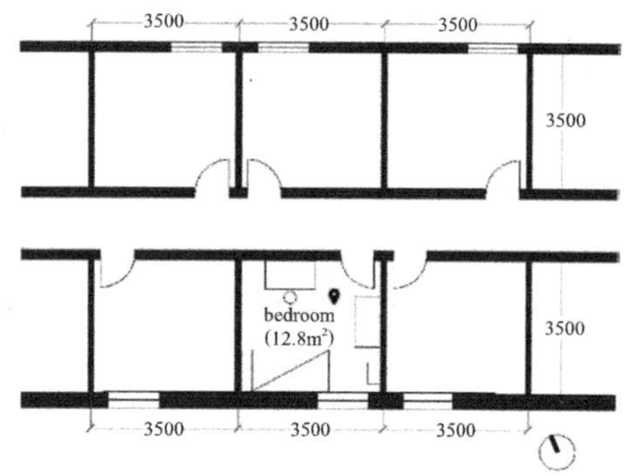

In the second year I moved to another room slightly smaller than the first one. In the third year the room was about $16m^2$ and two door-windows.

The most special area around the dormitory was a place that we called *el Pollete*, located in front of the dormitory's main gate and in between another dormitories. In this space there is a sort of long bench made of stone, L shape and one meter wide, which not only was witness about the history of the city but also about our long meetings with the female students of the adjacent dormitory.

The neighborhood where it is located the dormitory is called *Realejo* and it is an old Jewish sector of the city, with narrow streets, cobblestones and 5-storey building as the highest.I would like to remind again that I came from a city where the oldest building was 50 years old, and now I found myself living in a city with 500 years old buildings around me. In consequence, this place changed my inner feelings. Not only does it help me to connected with history, but also I began to appreciate the value of culture. Definitely, it is the most remarkable change environment of my life, to move from a small tourist city on the coast to a historic city nearby the mountains.

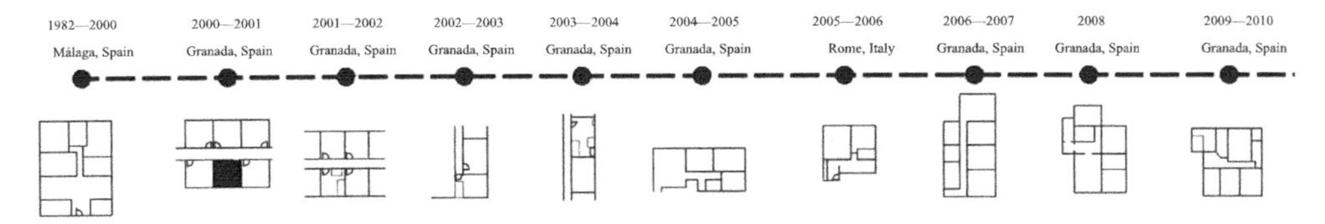

Rome (pian due)

In 2005 I received a scholarship to continue studying Architecture in a European country. So I moved to Rome to study for one year. And I moved with another two friends, whom I shared an apartment with. The apartment was located in a neighborhood called *La Magliana,* 5km south-west from the city center. It is a quite neighborhood where live families of native Italian workers and immigrants mostly from India.

Our apartment was on the second floor of a 6-storey red-brick building, around 60m^2 distributed of one living room, one room, one kitchen and one toilet. In this situation, we had to exchange the function of the living room into a room for two persons and keep the original room for one person. Every room has exterior windows faced to North and West, which doesn't benefit the entrance of so much sunlight, making the indoor a bit dark. A big wardrobe at the entrance was enough

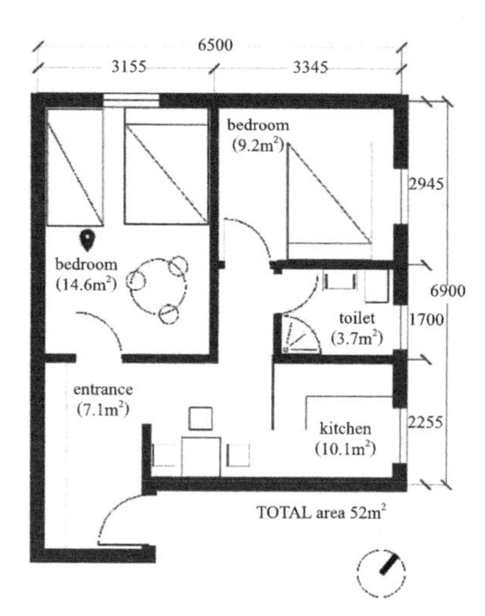

to storage all our clothes and belongings.

Although the house did not have big space and lot of sunlight, I remember every space in the room quite cozy. I think the reason is the house was well decorated by the owner. For example, the furniture were probably selected from Ikea, giving a sense of youth and happiness to the apartment. And one of living room walls were finish with stucco, enhancing the indoor light. The kitchen is small, around $10m^2$, but with the natural light from the window and youth furniture, it was a very cozy place to cook and eat. In one of the kitchen's corner there was a wood table for three to enjoy our meals.

Our neighborhood had nothing very special; just common facilities like barbershop, bank, coffeeshops and telephone booth regent by Indian immigrant.I don't remember to have a green park or area close to home to relax or play. Just a bike-path along the road of a couple kilometers length.

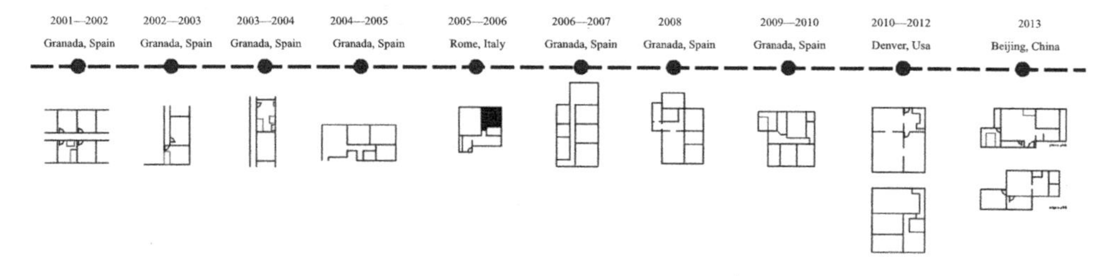

Denver (alcott)

In 2010 I was granted with a scholarship from the Spanish government to enjoy a one-year internship in a landscape studio. I was exciting, because it would be my first time living in USA, experiencing the American life that I used to watch in the TV.When I arrived to Denver, I did not have place to stay, so I stayed in my boss' house. After a few days, we moved to work at the Grand Canyon South Rim, so we stayed at the Grand Canyon motel for the next three weeks.
After a few weeks riding my bike and looking for an apartment, I finally found a perfect house. Rather than, the house found me.

The house is a gorgeous Potter Highlands Victorian house built at the end of the 19th century. It's situated on a hill northwest of the Platte River. When overlooking the city, it's the Highlands of Denver. The neighborhood is designated historic and it's called the Potters Highlands. Most of the houses are Victorian style built at the end of the 19th century, separated just a few meters one from each other and with a front yard and a backyard. Those houses do not have a garage, but it seems not a problem since you can park the car on the road in front of home. On the other hand, big trees provide shadows to the pedestrian who decided to walk or bike on the wide sidewalk. In general, I remembered abundant plants and vegetation everywhere in the neighborhood which made very pleasant to ride or walk around.
The main floor entrance faces to West and has a porch and a nice front yard with a few plants and vegetation. In this floor we had a small lobby, living room and dining room spatially connected to each other, and the kitchen opened to a cute backyard deck and patio. Tall ceilings were throughout main floor living and

dining rooms, with huge South and West facing windows for tons of natural light. Gleaming original hardwood floor are throughout. This home seamlessly blends vintage features, like original wood casing and decorative fireplace, with contemporary upgrades.

Upstairs are 3 bedrooms and 1 toilet. My bedroom was the middle one, about 14-15m^2 with two windows which face to East and South respectively. Although the room was not huge, the big amount of natural light that passed through those windows make the room feeling even bigger. The room also had a tiny dressing room just about 1.5m^2 but big enough for one person belongings.The walls were painted in dark green in an elegant combination with the hardwood flooring.

The owner used to live in the master bedroom, about 18m^2 with a large window face to West and a bigger dressing room. White walls painted and huge amount of natural light.With a double bed in the center of the room and facing to the window, it was the nicest room of all. Moreover, in this floor we had the small room, about 9-10m^2 facing to the East, and a bathroom with the shower, lavatory and toilet,

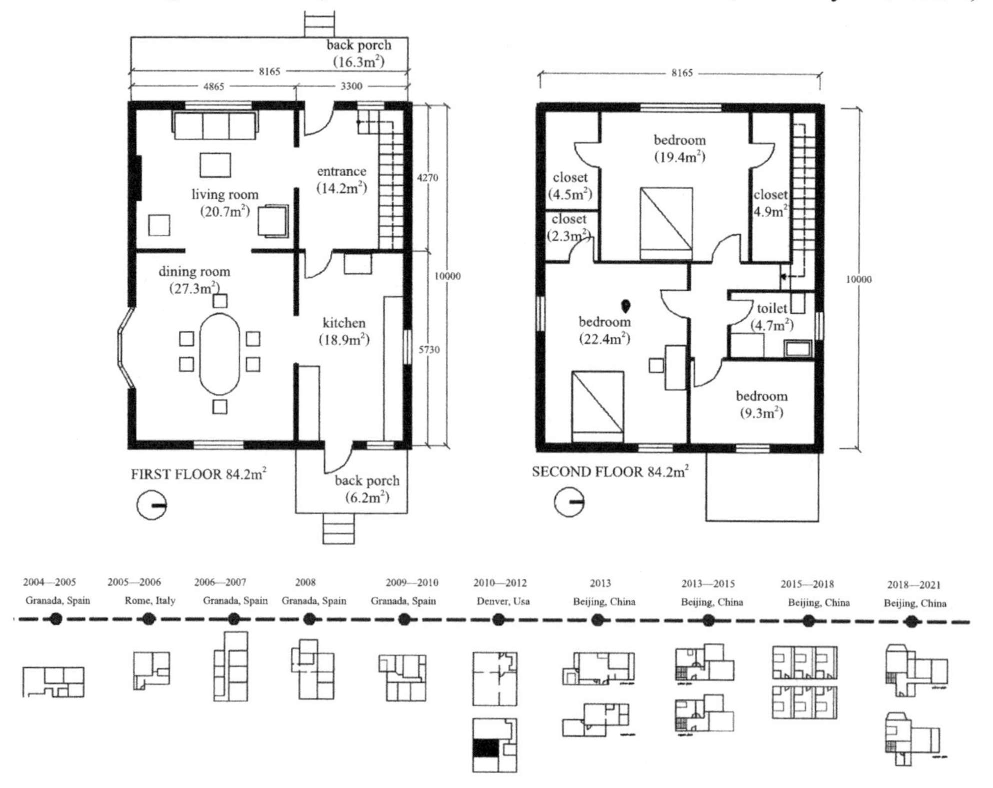

enough for both of us.

The basement is around 75m^2, and the perfect place to set the washer and dryer. There were no decoration or nice floor and wall, which was just finished in concrete and bricks respectively.Due to its soundproofing, I also used to come to the basement to play the guitar.

The backyard is 120m^2, with a big tree in the center and a few pine trees as a barrier to hide the wall division with the neighbors. This place was witness of our spring breakfast and summer party nights.

In total, more than 225m^2 of authentic paradise.

Beijing (Beiyuan, 798, Shuangqiao)

Beiyuan

In December 2012, I came to Beijing to work as an Architect. My first apartment was placed at the North of Beijing, close to the subway station Lishuiqiao Nan. It was on the top two floor of the building, the 16th and 17th floor of a 17 floor building. I didn't know at that time, but it is very common that the top two floors of a residential building in Beijing are used as a penthouse. In my case, this penthouse was around 120m^2 total area. In the first floor there was the kitchen, a toilet and 2 dormitories (the living room was converted into a room). And in the second floor there were another 2 dormitories, a toilet and one wide corridor that we used as a collective space to do sports.

The room I used to live was on the second floor and the smallest in the house (9-10m^2).Its walls were painted in pink and the furniture were white. According to the interior design of this room, I assumed that the previous user of this room was a

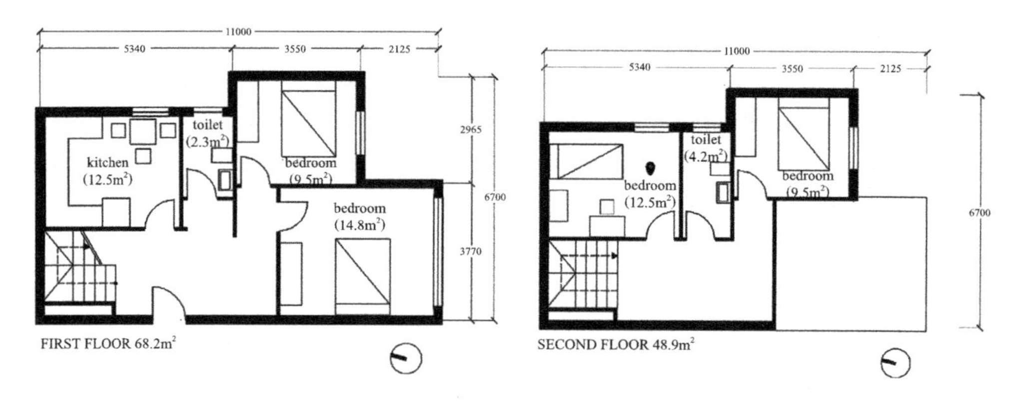

child. So, at my 30's, I suddenly became the child of this house.

Some negative aspect of living in this house was the lack of sharing area, such as the living room or the dining room. There was not such a place where we could meet and talk. The only area where we had the chance to meet each other was the kitchen, which was around 15-16m^2 with a square table (extendable) at the corner. I remember to spend long conversations with my roommates sitting at this table, eating and chatting with each other about our own hometowns. And also the kitchen was the only place in the whole penthouse where I could set a paella party, inviting roommates and friends to join it.

This house in Lishuiqiao provides me another situation that I had never lived before. To live in a living area. That was one of the biggest change comparing to living in Spain or Denver, where most of the houses and apartments were not set into a living area. It did not have anything special comparing to the other ones. Big square in the center where the elders did Taichi early in the morning, were with green areas, pavilion and even a small lake. Lovely in the morning and quiet at night. I used to have a walk after dinner with my roommates around the living area or even playing the guitar with my neighbors.

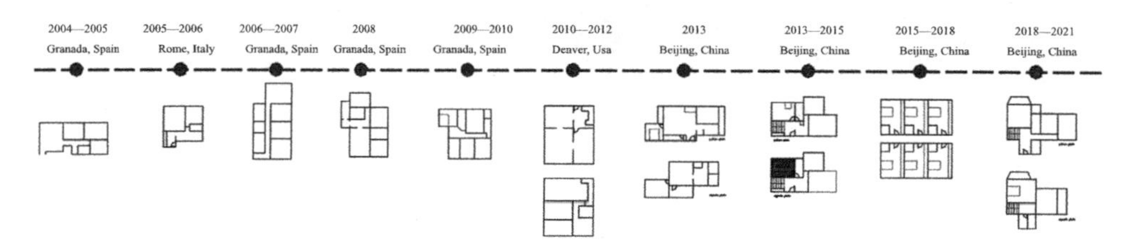

798

The company where I used to work decided to move to 798 area in 2015, in the north-east of Beijing. I also move to this area and lived there between 2015 and 2021.

During the first period I lived in a 6-storeys building dormitory, well located in the proximity of a park and just 10 minutes riding bike to the company. Although the location guaranteed a quite ambient, the inconvenient of this great location was the difficult connection with the transports, such as subway or bus.

In the 900m^2 first floor were the entrance, gym, shared kitchen and a large reading room. They were well decorated, with exquisite and refine furniture. Also it had so much natural light thanks to the glass façade of the building in the first floor. I remembered those spaces were quite large, spacious and well natural illuminated. It was a pity that a few years later every space were re-decorated and re-distributed to became a more efficient building.

We used to live in the sixth floor.I have to mentioned that every corridor to access to the room was very clean and prohibited to storage any furniture or personal item. I did like that feeling of arriving home just by walking through a clean corridor. The room was a double bedroom faced to the South, with a toilet, a kitchen and a desk within a total area of 24m^2. Through the window we could see the school in front of us, just a few meters to the South. Although the room was small, the amount of natural light and the peaceful environment made this apartment a comfortable place to live and study. One of the great advantages of living in this area was the convenient to go to the park and have a walk. Just a few minutes walking from the apartment and you find yourself in an enormous park surrounded by trees and lake.

I couldn't believe a place like this could exist in a city with 22 million residents. It was remarkable.

In 2018 and because of the redecoration of the building, we decided to move to another area closer to 798 Art District. The living area was of classic Chinese kind, with many green areas and parking underground.

Our house was the penthouse of the building, the 25th and 26th floors. Becauseit was the top of the building, we had the opportunity to enjoy great views of the city. The area in front of us was an office sector. Most of them buildings of one or two floor. So, no building could block our views of 798 and Wangjing Soho. I remembered taking many pictures of the great urban landscape we had in front of us. Offices, old chimneys, the sky, many were the objects to focus my camera on.

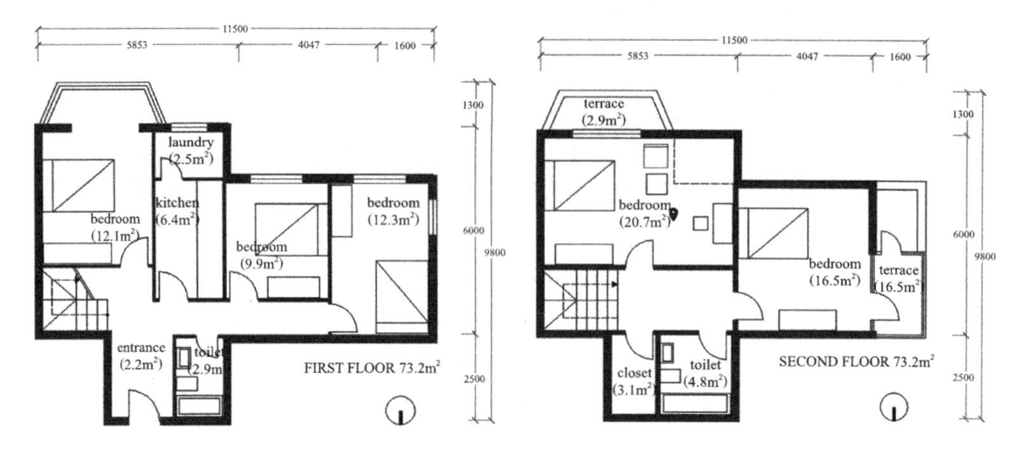

In the first floor of our penthouse we had three rooms, a kitchen and a toilet. Originally, the house had two rooms and one living room in the first floor, but the living room was changed into a room, to be economically more beneficial for the renter. The inconvenience is that this new design made the entrance of the house and the corridor a bit dark because of the lack of natural light. The kitchen was shared by everyone and sometimes I felt a bit uncomfortable because of the dimensions. Too small to be shared by 10 people, especially the fridge. The fridge was a disaster which made it sometimes even more difficult to find my food.

Upstairs we had two rooms, a closet and a toilet. Originally, our room were the living room and the owner decided to close it to be converted into a room. It was 19 square meters, with wood flooring, elegant wallpaper and a part sloped ceiling.

Some of the beams of this sloped ceiling were finished in wood. The biggest inconvenient of living in the top of the building was lack of thermal insulation, which made it very cold in the winter and very hot in the summer. Our room were faced to the South which guarantee big amount of natural light and warmer in the winter days. Anyways, occasionally we had to use the air conditioner.

The closet was a 8 square meters rectangular room and a big support to store all the personal items. The toilet in our floor was shared by just two rooms, which made it easier and more comfortable to use.

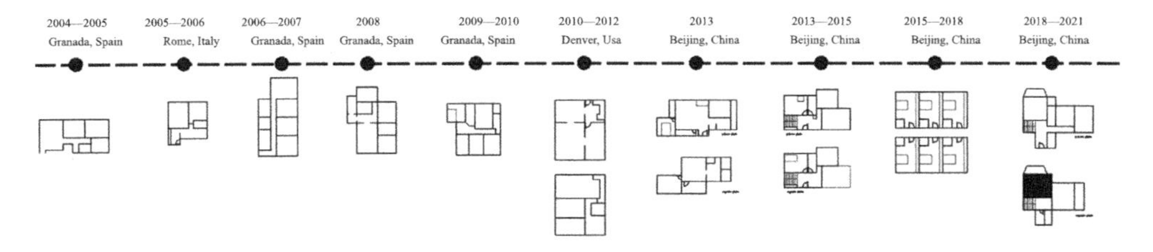

"Jinyu Kele"

In 2021, I had to move to Shuangqiao area because of work. I had the chance to work as an Architect in a company and we decided to move to be close to the workplace. So, since then I am renting an apartment close to Shuangqiao subway station, in an area called Jinyu Kele.

The most I liked this area was that I could feel the hustle bustle of the city. Our building sits along Shuangqiao Road, in the intersection with another road where the traffic and noise are constantly passing through. This is a kind of spectacle that I love to watch from our apartment.

There was no living area (which was something that I got used to from Spain) and the main entrance of the building was spacious and full of delivery machines. Our apartment is in the 12th floor of a 25-storey office building and the corridor to access to it. Although it is wide and well natural illuminated, it is being used to store some personal items from the neighbors (which is a practice that I cannot get used to, even though we practice as well).

The apartment is 76m^2 with wood flooring and high windows. There was one living room, double bedrooms, a kitchen and a toilet. The living room is the biggest room in the apartment and better illuminated. Its tall windows were faced to East and followed the curve of the building, which provided privileges panoramic views from inside. Like many apartments in Beijing, we had central heating which made the apartment warmer during the winter season. On the other hand, although the windows were tall and no curtains, we didn't feel very hot in the summer. The reason is that it faced to the East and we just received the warmer sun from the sunrise.

I used the living room as my own office, where I sat down on my desk and worked in the morning until lunch. Before lunch, I liked having a walk around the building for half an hour. Every street that surrounded the building were pedestrians, which made it more comfortable to walk or cycling. Also, some Chinese from other living areas liked to come to our community and played chess, ping pong or practiced in the exercise machines. We were very lucky to have all these public facilities around us.

Moreover, another issue that made us feel very lucky about this location, was the convenience to find canteens, supermarket, barbershop or bakery, which were well located in the first floor of the building.

Although I felt quite satisfied living in this apartment, there were a few aspects that needed to be improved to reach a better-quality live standard.

The first one was related to the human behavior. Indoor, many people used the corridor as a storage for personal items or trash, making the public area with bad smelling, dirty and visually poor. The same happened in another public area, the outdoor pedestrian street where many people did not pay attention to public hygiene.

The second one would be related to the maintenance. For example, a small budget was used to paint the wall corridor, fixed skirting board or maintain clean the elevator. These datils of the building, although not essential to live, helped to make the life of the residents more comfortable and reach a sense of peace. And the same happened to the outdoor facilities in our community, such as benches, floor or vegetation, sometimes dirty or broken.

The last one would be related to the circulation of some private motorbikes on the pedestrian road. This kind of circulation needed to be restricted and regulated, not only for the security of the pedestrians, many times the elderly and kids, but also for the maintenance of the public space. Due to the intensity of this circulation, many paving were broken and vegetation destroyed.

Changes Never Happen Overnight

by Kang Mengliu and Kang Bin'an

Born in Erguanzhai Village, Shengjiaba County, Enshi City, Hubei Province in the 1970s, I am currently employed by a company. Approaching the age of fifty, I have never taken a close look at my hometown. My daughter expressed her desire to learn more about our hometown and share its story with others. Therefore, she encouraged me to participate in this endeavor. It seems like a good idea to reminisce and summarize the transformation of our hometown with young people from the generation of 2000s. The transformation in my hometown not only reflects the changing times but also captures the evolving relationship between humans and nature. From various perspectives, I would like to present the development and changes that have occurred over the past 20 years.

The first issue I would like to address is the living environment. Over the past two decades of development, my hometown has undergone significant changes. Erguanzhai Village, which used to be a place where humans and animals coexisted with scattered livestock, careless garbage disposal, piled up firewood and hay everywhere, and freely flowing sewage, has now become a national 3A-level scenic spot with clean rivers, pleasant living conditions, leveled roads, and beautiful scenery. The construction of roads has significantly shortened travel time back home. I believe this is the best change for my hometown.

Since it is a scenic spot, the environmental requirements are much stricter than in ordinary areas. For example, poultry raising has been banned for many years, and garbage classification and centralized management have been implemented. Roadside garbage is regularly cleaned along with the maintenance of river hygiene. Additionally, deforestation is prohibited to maintain the forests coverage levels.

In recent years, with government support, many fruit trees have been planted along the roads for people to appreciate by taking pictures or picking fruits as souvenirs or to quench their thirst in summer. Various flowers such as peach blossoms, rhododendron flowers, camellia flowers, and cherry blossoms have also been planted to embellish the surrounding environment. Villagers are encouraged and supported in growing crops and selling fresh agricultural products without pesticides spraying during the growing season.

Scenic Spot Introduction Plaque

Cherry Blossoms Season

In addition to changes in the natural environment, it is also essential to consider the rural construction and maintenance. Do you still remember the old mud houses we used to live in when we were young? They would often shed tiles and dust, which would be blown around by the wind and cause dust in the surroundings. Nowadays, beautiful wooden houses with elegant exteriors and spacious residential areas no longerexert pressure on the natural environment.

Panoramic View

Compared to my daughter, I pay particular attention to the changes in roads. The muddy roads that used to dirty shoes and socks when walked on have also gradually evolved into sandy paths and pebbled ones, further transitioning into gravel and cement ones. They now represent smooth, comfortable, and visually appealing asphalt roads. Even the previously inaccessible routes leading to people's homes have now become well-connected avenues.

Whenever I return home, I marvel at how good our roads are! The roads now reach our doorsteps directly. Unlike before, they were always muddy, dirty and smelly, long but unable to extend all the way to people's homes. Walking on them was not only time-consuming but also labor-intensive.

The young people, including my daughter, would give us a vague and indifferent expression when we talked about such things, which always bothered me. She was still very young at the time and couldn't understand why we kept saying those things. It wasn't until she had the chance to take a shortcut with her paternal grandfather by climbing a mountain to visit her maternal grandmother's house that she got to know what we were marveling about and why we endeavored to struggle in large cities. Her understanding became comforting for me.

Photos of the Roads

The stone bridge of the past has become a witness to time, gradually fading out of people's lives and becoming a well-known scenery that evokes nostalgia. Its new partner, the scenic bridge, carries people's new hopes and has become a must-visit spot and the best viewing place for visitors. Looking from afar, the two bridges form a striking contrast, serving as an effective reference for reflecting the enormous changes around them. I still remember when my daughter fell off the old bridge and lost a tooth when she was young. She took it very seriously and said, "I will never walk on this broken bridge again." However, every time we go back I see her walking around the old bridge. What is she thinking? I have never asked.

Photos of the Ancient Stone Bridge and the Newly Built Scenic Bridge

My daughter told me that the environment in Erguanzhai Village today can be described as having a clearer sky, cleaner water, and fresher air.

Sunset Clouds

Rivers

The second issue I would like to discuss is the support and assurance provided by environmental protection policies. Without government support and financial aids for infrastructure construction, the development of our small village may still lack direction. Before being certified as a 3A-level scenic spot in 2016, we had already observed that our hometown was undergoing transformation. My uncle, who was then an village official, commented, "The senior administration has issued a policy to develop tourism in Enshi City, so we should also make an effort. It would be wonderful if we could be certified as a scenic spot." From then on, many individuals who had migrated to large cities for work returned to renovate their houses and decorate their courtyards, contributing to the development of their hometown. The local government started allocating funds for road and bridge construction as well as trees planting. As anticipated, we successfully made it onto the list of scenic spots. Weak and dispersed forces alone can hardly achieve significant progress and development; instead, strong and effective policies and mechanisms truly contribute towards achieving goals.

Local Residents Refurbishing Their Houses

Daily Life

Thanks to the policy guarantee, people have the opportunity to fully utilize their skills and make meaningful contributions. Every time I return to my hometown, I always hear the elderly villagers talking about recent news: how the township government is providing subsidies for building houses in order to offer better accommodations for tourists. This fills us with hope for a brighter future. Actually, my wife and I plan to run an agritainment restaurant back home after retirement as a way of supporting ourselves while enjoying our later years without excessive strain. This will enable us to live independently and peacefully within our hometown.

Elderly villagers talk about planting more trees to purify the air, provide a better visual effect, and create pleasant memories for visitors. They also mention that as elderly people, they now have activities to engage in such as dancing, participating in performances, and playing games with young people. They no longer feel as bored as before. It appear that our days are filled with hope. Every time we look at the photos and videos of grandparents taken by the village committee we feel particularly happy and often exclaim, "Grandmas look younger and more energetic than before."

Bonfire Celebration

Guessing Lantern Riddles

They mention that in the past, after the traditional Lantern Festival ended, young people would all leave. Now there is a Lantern Festival party held in the village every year, which invites the younger generation to participate. Although they don't fully understand the meaning of these activities, it's great for them to finally be able to celebrate the Lantern Festival with their elders. In fact, it is not the generation of 2000s who enjoy the most fun at the Lantern Festival, instead, it is their parents, uncles and aunts who help organize the events and engage in the village construction efforts. We regret that we couldn't stay for longer due to working elsewhere. What a shame that we haven't attended the party yet! However, every time when we watch our familiar members participating in these activities and winning prizes and money through video calls, we feel exceptionally happy. I have also considered moving back to my hometown some day when we have saved some money while our children are getting older and more development opportunities

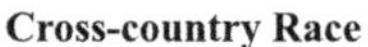

Cross-country Race

Lighting Candle Activity

become available.

They mention that from now on, all trash will be collected and disposed of in designated places, and no one is allowed to throw it around carelessly anymore. Public trashcans will be installed everywhere on the roads so that people no longer have to carry waste in their pockets while walking.Villagers no longer have to guard their forests in the mountains since the local government and the village committee provide safeguard support against tree thefts, allowing villagers to enjoy more leisure time. They can go swimming, fishing, and take a break by the river which has regained its vitality and flows gently again... One after another these statements are made by the elderly, reflecting the effective efforts made by our local government in environmental protection.

Every time my daughter heard these comments, she would express her admiration. I believe that only when these policies and measures have truly been implemented can they be vividly discussed by the elderly villagers who are the most authentic witnesses there and who do not know flattery, embellishment or exaggeration like those chaotic pieces of information from the Internet; instead, they only understand what they have seen with their own eyes. With limited education, I may not fully comprehend what my daughter and her friends discuss. However, from her expression, I can still grasp her approval of the changes of our hometown in recent years. Indeed, I am also exceptionally happy about it as it is the place where I have lived for decades and will return to in the future.

Finally, I would also like to share my suggestions for the development of my hometown. Currently, our local tourist attactions are flourishing and we believe they will continue to thrive in the future. However, I think there are still some issues that need to be addressed further.

If self-discipline is not sufficient, we need to enhance our regulatory efforts. Every year, I take my family back to hometown either to escap the summer heat in cities or for winter holidays. We stroll around the town and countryside, and during our discussion with the younger generation, we have identified the following issues:

a. Despite having public trashcans, garbage can still be seen lettered on the roads, in rivers, by the riverside, and in parking lots.

b. Although there are many public fruit trees that can be picked, they have already suffered from damaged branches and fallen fruits which hinders their growth next year.

c. Numerous public facilities have been damaged, including scratches on the newly built corridor bridge, damaged trashcans, destroyed graphite decorations, missing paintings carved on rocks meant for visitors' enjoyment, and shared electric vehicles being left randomly.

d. Crops in fields are being trampled and stolen while visitors enter private farms without permission.

These situations serve as a warning that we should improve regulatory work while developing tourism in order to effectiely maintain order and protect the environment, thus saving a huge amount of unnecessary repair and reconstruction costs. I believe implementing such a win-win measure could yield positive results.

Environmental governance is of great importance, with a focus on identifying the sources of pollution. At first glance, progress has been achieved in environmental protection, with substantial efforts invested. However, the pollution that occurs during this process poses unforeseen consequences that are difficult to measure or control. For example, in my memory, the national highway from our Shengjiaba Township to Xianfeng County was planned and started in 2020; so far, it has been divided into several sections and some remain unfinished. As a result, on the main road passing through the ancient Juipu village and Xiaoxi village, sand and dust fly wildly on sunny days while it becomes muddy and inaccessible during rainy days. We have to pass through this section when we travel back home and all we see is sand and dust floating in the air. We have to close the windows of our car which is already covered by sand and mud. We experience turbulence along with concerns about the environmental situation as well as inefficiency in addressing these issues. Unfortunately, we can offer little help except for anxiety: what impact have these roads had on air quality over so many years ? Does this damage outweigh all our protective measures and efforts? Nevertheless, we need to change our mindset—if we can eliminate pollution from the beginning, will we still have to spend significant time dealing with it later? In short, cutting off pollution at its origin is crucial.

We should always be prepared for potential problems, and environmental governance and protection require a long-term commitment. As our region already has a better environment compared to others, we should focus on implementing our protection plans more efficiently. however, I realize that my hometown lacks sufficient population as many young people leave to work elsewhere, leaving mostly the elderly behind who may struggle to fully understand the importance of environmental protection or how to take action. This makes it difficult to implement long-term measures. Therefore, we need to make future plans and attract young people back to their hometowns. At the same time, the local government should recruit talented individuals to advise on better policies and ensure their implementation. To achieve this, provincial and municipal governments should issue policies encouraging young people's return home and gradually implement these policies at the county and village levels with strict monitoring of each step along with providing benefits, subsidies and support for individuals. After all, people work for making a living; if there are well-paying job opportunities nearby, they would be very happy to come back, wouldn't they? I firmly believe that if we can successfully solve the population problem, both environmental protection and development will greatly accelerate.

For the future development direction of our hometown, my daughter believes it should encompass a comprehensive solution involving tourism, special agriculture, and forestry. Meanwhile, I believe that in order to achieve prosperity, we must prioritize road construction of roads. The establishment and improvement of infrastructure serve as the foundation for any future development. Therefore, it is evident that our greatest efforts should be directed towards designing, planning, renovating, and maintaining of the regional environment while also addressing environmental problems. We believe that in the future, our hometown will become a livable rural area with both beautiful natural environment and an enchanting humanistic atmosphere.

Tale of the Dazzling Beauty in the Northern City

by Lan Dong and Chen Haiyan

In 1976, I was born in Harbin, a city full of poetry and history. At that time, Harbin resembled a gentle beauty, retaining the charm of Russian architecture while gradually exhibiting signs of vigorous vitality in the early stages of reform and opening up. This city used its unique style to weave warm and profound memories for me and my child.

Harbin, located in Heilongjiang Province, China, is known as the "Oriental Petit Paris". Its beauty is not only reflected in the historical buildings that have withstood the test of time but also displayed in the ever-changing modern urban landscape. The living environment in Harbin resembles a moving painting, skillfully integrating nature and culture.

On the banks of the Songhua River, tall trees provide shades, a gentle breeze blows, and the water is calm. In summer, the sunlight shines on the river, creating ripples of golden light as if countless elves are dancing and leaping in joy. Children play by the river, running barefoot on the sand, chasing waves, and enjoying the gifts of sunshine and nature. That innocence and joy are still cherished memories in my heart. As night falls, lights illuminate both sides of the street with dazzling neon colors. The European-style buildings on Central Street seem to transport people to a foreign country, attracting numerous tourists to stop and admire them. In the heart of the city lies a bustling scene where historical buildings coexist with modern skyscrapers that shine together, narrating tales about its splendor and transformation. Modern commercial districts pulsate with boundless vitality and creativity within this vibrant city's veins. Towering skyscrapers reach for clouds while glass facades gleam brightly under sunlight like dazzling mirrors reflecting urban hustle-bustle and vitality.

The beauty of Harbin's living environment cannot be fully described in words. It is so profound and elegant that it not only possesses external splendor and prosperity but also encompasses internal cultural depth and humanistic care. This city utilizes its wisdom, efforts, warmth, and care to provide Harbin citizens with a sense of peace and warmth at home.

Saint Sophia Cathedral in Harbin

As the years passed, I grew from a child who ran along the riverbank to a parent. My child continues her childhood and growth on the land of Harbin.

My parents, who were bank employees before they retired, got caught up in the "house allocation frenzy" of that era and were allocated a 90-square-meter residence. Stepping into the house, every corner exuded warmth and comfort of a home. The spacious living room was furnished with soft sofas and wooden coffee tables, while the interior decoration was simple yet elegant. White walls adorned with family photos added a warm and cozy atmosphere. The bedding in the bedrooms was neat and tidy, and the floors shone like a mirror. It seemed that every corner of the house had been meticulously maintained by its owner. The kitchen and bathroom were equally clean and tidy, with all facilities and utensils properly arranged. Sunlight filtered through curtains, casting shadows on smooth floors, creating an ambiance of tranquility and comfortable lifestyle. Every corner of this house revealed the owner's love for life as well as their appreciation for their home.

Indoor Environment of My Parents' House

Every day after school, my daughter would run to the school gate to catch the school bus and rush home to enjoy delicious dishes prepared by my mother. She would share school stories with her grandparents. When I finished work, we would take a walk together to the nearby night market, where the hustle and bustle of people's everyday life always made me feel connected. The diverse crafts on the street stalls, the aroma of grilled squids on the iron plates, and the refreshingly

My Parents' Residential Building

chillness of watermelons were all flavors that no child could ever forget from their childhood memories. If my daughter did well at school, I would take her on a bus ride to Central Street for a leisurely stroll. There, we could listen to beautiful melodies at Harbin Summer Concert on the gold-worthy "Bread-Stone" paving avenue or watch Austrian musicians playing the violin in century-old western-style building with elegant postures. Occasionally, I might even ask a street artist to draw her a portrait or savor authentic Russian cuisine at the famous Modern Restaurant.

Residents of Harbin by the Songhua River

Central Street at Night

The Gold-worthy "Bread-Stone" Paving Ground

Victory Monument Against Flood

Pedestrian Walkway of Central Avenue

Later, due to work reasons, we moved to the Xiangfang District where my husband's parents lived, which presented a completely different appearance from the Daoli District where we used to live. The Xiangfang District used to host large equipment industrial factories such as the Harbin Electric Machinery Plant, the Harbin Steam Turbine Plant, and the Harbin Boiler Plant, leaving traces of the industrialization era. The cityscape was more complex and diverse with red brick houses, the black and thick electric wires wrapped around concrete pillars, and the dense shops on the commercial street—all forming unique memories for me and my child. In the afternoons, three generations of us often went for walks in the nearby botanical garden, basking in the warm sunshine while feeling tranquility and harmony of nature. Swans frequently swam leisurely on the lake in the park with their graceful reflections swaying in water. Willow trees by the lake gently swayed in breeze. Walking between pavilions and towers allowed us to witness energetic elderly people dancing in pairs or attending concerts featuring Chinese erhu (traditional two-string violin) and accordion—a harmonious fusion of Eastern and Western music; or enjoying scenes of retired workers' leisure life like feeding birds, admiring flowers, or fishing.

The Botanic Park near Granny's Home in Spring

Other Parks near Grannys Home in Autumn

Despite the regional differences, the attitude towards life among Harbin residents in those days was remarkably similar. They walked briskly yet with an air of composure, as if they were in control of the rhythm of life. Their faces always carried a gentle smile, reflecting their love for life and yearning for the future.

As my child grew up, I witnessed the remarkable changes in education and living environment in Harbin and throughout Northeast China.

When my child started elementary school, we bought a house in the Zhaolin Primary School district to provide her with a better education. At that time, the real estate boom was sweeping the country, and house prices were soaring like wild horses. Especially the houses in the school district were in high demand and sold at exorbitant prices.

Despite its small size of only a few dozen square meters, our house was brimming with our hopes for our child's future. Every inch of space was meticulously planned and fully utilized, as if the air itself exuded compactness and fullness. However, the limited space also meant a crowded life. The room layout was tight and confined, with furniture pressed against the walls, leaving little extra room for movement. The kitchen and dining room were combined, and a small dining table occupied almost all of it; hence, when we dined, we had to huddle together in the confined space. The bedroom was even tinier; it could barely accommodate a bed and a simple wardrobe, making maneuvering around quite challenging. The family's work and study supplies had to be piled up in one corner, making the already cramped space even more unbearable.

Furthermore, the outdoor environment was also difficult to accept: rubbish was littered everywhere due to a lack of proper community management, while unpleasant smells permeated through the air. Small vendors set up unregulated stalls that block motor vehicle lanes without reason, thus obstructing the only entrance and exit of the road. The dilapidated buildings on both sides of the street had peeling walls. Green belts appeared desolate, lacking any sign of greenery. Northeast China, a once prosperous land, was gradually declining in the tide of history.

Five years later, my daughter enrolled in Harbin No. 76 Middle School and embarked on her middle school journey. Due to limited land resources in central urban area, the school had to relocate three times within just four years. The

Indoor Environment of the School District Apartment

Cramped "Old and Run-Down" Buildings

classroom facilities used to be outdated with worn-out desks and chairs; sanitation conditions were also concerning, with irregular garbage disposal and sights of unsanitary corners. These problems made me acutely aware of the weaknesses in Northeast China's school infrastructure and the lack of sanitation and service guarantees. Later, as I saw my child diligently studying in the new campus, I felt a sense of relief mixed with worries because I myself was facing double pressure from work and family, experiencing firsthand life's hardships and responsibilities weighing on my shoulders.

Due to my job, my daughter occasionally visited Harbin's Pingfang District. Although it might not be as prosperous as the newer districts, it possessed its own unique charm rooted in historical heritage. In large factories like Harbin Aircraft Industry Group of AVIC (Aviation Industry Corporation of China) and Harbin Dongan Auto Engine Corporation, one could witness scenes reminiscent of old movies from the 19th century: swarms of workers dressed in blue uniforms pouring out through factory gates. However, their expressions no longer radiated joy and pride but rather exhaustion and helplessness caused by life's burdens.

At that time, Northeast China seemed to have been enveloped in a thick layer of gloom. The once vibrant and energetic industrial atmosphere had gradually faded with time. Walking through the streets and alleys, one could sense a profound feeling of loss. People's eyes had lost their sparkle and gained a touch of confusion. Perhaps they were contemplating how to face the sudden changes or search for new paths. But whatever the case maybe, the once grand ambition appeared to have vanished.

Harbin's No. 76 Middle School

After the middle school entrance examination, my child was admitted to Harbin No. 3 Middle School (Qunli Campus).That was not only a crucial period for her personal growth but also an important moment for the changing living environment in Harbin, especially in the two new districts.

Qunli New Area, located in the new heart of the city, brought us endless surprises and joy. Here, there were no more dilapidated and dreary pasts; instead, a brand new and lively land emerged with numerous large buildings that had sprung up like bamboo shoots after the rain—narrating the story of the new era through their unique postures.

In order to provide our child with a better learning environment and living conditions, my husband and I screened numerous real estate companies and finally decided to purchase a residential apartment in Shenghe Century Community.

Upon entering the apartment, you are greeted by a spacious and bright living room. The wooden floor is smooth and warm, both environmentally friendly and durable, complementing the light yellow walls to create a harmonious atmosphere. It emits a natural glow, as if it can absorb the warmth of the sun and slowly release it throughout the space, creating a cozy and comfortable living atmosphere.

In the center of the living room, a row of beige sofas quietly stands, featuring smooth lines and soft fabrics. Sitting on them makes one feel as if the whole body is gently wrapped, unconsciously inducing relaxation and enjoyment of peace and tranquility painting hangs on wall behind the sofa with vivid colors and delicate strokes, adding an artistic touch to space. Beside the sofa stands a simple yet elegant TV cabinet that is stable while still maintaining fashionability. On top of this cabinet are neatly arranged family souvenirs from travels; each carrying beautiful memories. The family photos from different stages hold even more precious value as they document our family's journey through ups and downs along with moments of laughter. In addition, there are some calligraphy works created by our close friends during their leisure time, which silently convey our family's attitude towards life.

On one side of the living room, several large windows bring the scenery of the community into the room. The sunshine streaming in through the windows falls on the floor, creating a pattern of dappled light. Outside, a red running track spreads in the sun, serving as a must-go route for morning joggers. In the morning, you can often see residents sweating it out on the track while enjoying the joy of exercise.

On the other side, the outside view from the window is even more expansive. Through it, you can witness the first rays of sunshine illuminating the earth in the early morning, the hustle and bustle of vehicles on Yangmingtan Bridge, and orange hues of sunset reftected on the riverbanks. The rise and set of the sun are like masterpieces of nature—warm and grand. When dawn breaks, sunlight is soft and full with hope that infuses a new day with vitality and energy. As the sun sets in the evening, sunlight gradually fades away, leaving a glowing afterglow that cloaks the earth in a serene hue.

Additionally, some potted plants are placed in the corners of the living room.

Living Room

Dining Room (Riverbank View)

Whether they are large green plants in the corner or the small potted ones on the windowsill, they all exude vitality and vigor, making people feel the beauty and vitality of life while adding a touch of natural charm to the room.

The Music Corridor in Qunli New Area, resembling a long ribbon, shines brightly in the summer sun and becomes a dreamy shooting location for wedding photos,

Daughter's Bedroom

Panoramic View of the Community

Landscaping in the Community

Night View of the Community

Main Road near the Community

leaving beautiful memories for every couple. In winter, it transforms into the home of a giant snowman that attracts countless tourists to come and take photos, leaving their footprints behind.

Commercial giants such as Wangfujing Shopping Center, Intime Department Store, and Yuan Da Shopping Center have all established their presence here, offering a wide range of products and exceptional services to attract numerous young and middle-aged individuals who seek the pleasure of shopping and wish to stay in tune with fashion trends. Meanwhile, parks like the Sports Park and the Music Fountain boast abundant greenery, providing ideal spots for leisurely strolls or weekend getaways. Walking along the shaded pathways while listening to birdsong and immersing ourselves in nature's embrace transports us to a paradise far away from the mundane world; this experience leaves us feeling refreshed and rejuvenated.

The Music Gallery and the Giant Snowman in Winter

In addition to the excellent facilities for business and cultural activities, transportation in Qunli New Area is also becoming increasingly convenient. Subways, buses and other means of transportation intertwine to form a network, making it convenient and comfortable to reach both the city center and other area. The 15-minute business circle coverage in Qunli New Area has greatly enhanced our daily lives by providing us with convenience in shopping, dining, and entertainment within a short distance. This efficient and convenient lifestyle has attracted a large number of young and middle-aged talents to settle here, making

Qunli New Area a vibrant and creative community.

The government's focus on relocating and resettling rural residents has also ensured that former residents have been properly settled, vividly demonstrating the successful integration of urban and rural development. Furthermore, government departments that located in the old city center have gradually moved to the new area, further enhancing Qunli New Area's urban functions. In terms of educational resources, Qunli New Area has spared no effort. Alongside the prestigious Harbin No. 3 Middle School, the government has established many branches of local key middle and high schools and introduced numerous excellent educational institutions, allowing Harbin's youth to enjoy a more diverse and high-quality education here.

The security situation in the Qunli New Area is even more reassuring. Security guards are on duty 24 hours everyday in the residential area, strictly managing property entry and exit, while police cars patrol the main streets every evening. The

The Music Gallery Overlooking the Community Across the Street

Birds in the Parks

The Artificial River within the Park

Wide Motorway with Eight Lanes

robust security system and professional security team provide residents with a safe and harmonious living environment.

Songbei New District, where modernity meets classic elegance and bustling coexists with tranquility, is quietly nestled in the embrace of Harbin like a poem, a painting, and a dream. Every weekend or during holiday, I bring my child here to learn about the changes and development of Harbin.

In Westred Square, you can truly sense the passage of time while appreciating the charm of creativity. The red brick walls and old factories tell stories of the past, while modern cafes and art galleries showcase new vitality and energy. The weight of history collides with the trend of modern times, igniting boundless possibilities.

At Wanda Plaza, you can savor delicacies from around the world and enjoy a gastronomic feast; simultaneously experiencing sheer joy through shopping. Harbin Grand Theater, with its unique architectural style and outstanding artistic quality, has become a new cultural landmark in Harbin. Here, you can enjoy world-class concerts, theater performances, and dance shows while immersing yourself in the limitless allure of art.

Furthermore, the theater frequently hosts various art exhibitions and cultural activities that provide an ideal platform for citizens to exchange ideas and thoughts. Whether gathering with like-minded friends to discuss secrets of art or enjoying an exhibition alone seeking moments of peace and serenity—this venue offers it all.

As a parent born in Harbin after 1975, I cannot only witness the city's history and present but also envision its promising future. I believe that in the days to come, Harbin will continue to attract more individuals with its distinctive allure, enticing

Harbin Grand Theater

them to settle down, pursue their careers, and flourish here. My family and I are committed to writing our our own stories on this remarkable land.

From cramped and narrow school district apartments to spacious and bright commercial apartments, from chaotic and disorganized community to tidy and pleasant living spaces, every step of the transformation has witnessed a significant improvement in the living environment of Harbin City. This is not just a change in physical structures but rather a profound enhancement in living quality and a sense of belonging.

The deep concern and firm support from the state and the government resemble an invisible yet powerful guiding force, leading the city towards a brighter future. Every policy and plan issued by the government reflects the wisdom and collective efforts of society, injecting boundless vitality into the dynamic development of the city. However, it is also important to acknowledge the challenges and shortcomings in improving people's living environment in Harbin City.

In future development, we still need to address issues such as the rationality of urban planning, the urgency of environmental protection, and the completeness of public service facilities. We wonld continue to adhere to the principle of scientific development, pay attention to ecological balance and sustainable development, deeply explore the inherent laws of urban development, and guide every decision and plan to be more people-oriented and aligned with reality based on their needs and expectations.

More importantly, we should prioritize ecological balance and sustainable development, striving to promote economic growth in urban areas while safeguarding our pristine mountains and clear waters. In this way, the city can coexist harmoniously with nature, allowing Harbin to continue flourishing with renewed vitality and become a future haven that people yearn for.

Ode of the Era Impressions of—Sixty-Years of Human Settlement

by Li Hongqin

I was born in Yantai City, Shandong Province in the early 1960s of the 20th century. At that time, Yantai was still a very small city. The place where we lived was quite close to the former Yantai Laishan Airport. My memories of Yantai City are of a house standing near a small river during my childhood, which had clear and clean water with sufficient volume. The social development was relatively backward; most people lived in poverty. Daily life was not very convenient; for instance, there was no tap water at home, and all toilets were public restrooms. However, the surrounding environment was beautiful with abundant water resources and lush vegetation.

In 1966, I moved with my parents to Suizhong County of Huludao City, Liaoning Province, where the social development was much less developed than that in Yantai City.

Firstly, the roads. While traveling by bus from Shandong Province to Liaoning Province through Hebei Province, I noticed that roads in Qinhuangdao City of Hebei Province were made of asphalt; however, after crossing the boarder between Hebei and Liaoning Province, they turned into sandy earth roads.

Secondly, the commodities. The variety of commodities available there was quite limited; however,their prices were relatively low. People rarely dined out at restaurants; instead, they bought food products such as the deep-fried dough sticks that were transported from Shandong Province and Beijing City, which locals referred to as "oily cakes". During the day, people consumed coarse grains like sorghum, sorghum flour, and corn flour. The taste of the sorghum rice made from sorghum and the steamed buns made from sorghum flour was very rough. In contrast to Yantai City, there was always a shortage of wheat flour; even with "grain coupons", it remained difficult to purchase wheat flour.

Thirdly, the houses. The houses were mostly very old. In fact, there were more old houses in northeastern China compared to other regions, and new houses were rare there. The climate in Suizhong County was completely different from that in Yantai City. In Yantai, although it also snowed in winter, the lowest temperature usually hovered around -6℃ or -7℃ . However, winters here were much colder with heavy snowfall, and the lowest temperature could reach -20℃ . During winter, there were very few vegetables available, so people had to dig cellars to store them beforehand. At that time, there were many Japanese-style buildings in the County and some daily items also bore marks of the Japanese colonial period; for example, chairs at barber shops originated from Japan. As a child who wasn't tall enough yet, I had to have a wooden board placed on the chair so that I could step on it before I sitting down for a haircut.

Fourthly, urban planning. There was no urban planning in the County, nor the bus routes. The houses were built chaotically without an overall layout plan. When standing at the train station and looking southward, we could see endless fields stretching to the edge of the sea; however, the sea itself couldn't be observed from the train station. To the north of the train station was the county center, where a civil arterial road ran parallel with the train tracks that was oriented east-west. Public service facilities such as the county government, hospital, and school were distributed near this main road. These were my impressions of Suizhang at that time.

In 1970, we moved to Qinhuangdao City in Hebei Province. Qinhuangdao is an ancient city boasting significant strategic importance from ancient times to the present. At that time, although there were no local buses available within the central area of the City (now known as "Haigang District"), long-distant bus routes operated to connect different districts of the City, including a route directly linking the city center with Shanhaiguan District. We resided in Shanhaiguan District, famous for its reputation as "the First Pass of China". However, what many people may not know is that the Eight-Nation Alliance once stationed troops in Shanhaiguan and established camps. *(Translator's note: The Eight-Nation Alliance refers to the coalition of eight countries-Britain, the United States, Germany, France,*

Russia, Japan, Italy, and Austria-Hungary - that invaded the Qing Dynasty in China in 1900.)

The Eight-Nation Alliance Camp Site is still well-preserved and is considered the largest and most complete one of its kind in China. The old camp site, located southwest of the Old Dragon Head Scenic Area, was actually occupied by British troops; therefore, several dozen British-style villas used to exit in Shanhaiguan near the sea. These villas had simple architectural styles and were all equipped with red brick pitched roofs. Furthermore, great attention was paid to creating a green environment around each villa by surrounding them with plane trees. There were also many British-style villas further away from the sea. However, these buildings were almost all demolished during the 1980s and 1990s due to construction needs for restoring the supporting facilities and infrastructure for the Shanhaiguan Great Wall.

These houses were designed and built to a very high standard, with their foundations raised half a meter to one meter or more above ground level in order to prevent rainwater from flowing back into them. Similarly, in Qingdao City, there were many German-style buildings along the coast that were also constructed in this manner; their foundations were elevated more than one meter above ground level, making them highly resistant to flooding. In contrast, local houses in the Yangtze River and Yellow River basins had much lower foundations and were often affected by heavy rainfall. Apart from their excellent designs, these buildings also boasted high-quality construction. I am not sure of the roofing material used; however, when many of these structures were demolished, people took the metal sheets from the roofs and created large basins for washing clothes and bathing children. Despite frequent exposure to water, these basins showed no signs of rust at all.

After working, I relocated from Qinhuangdao to Cangzhou City in Hebei Province. When it comes to coastal cities, most people would think of Dalian and Qingdao first. In terms of Hebei Province, people may think of Qinhuangdao and Tangshan. However, Cangzhou City has also been located adjacent to the ocean since ancient times and was once the starting point of the northern "Maritime Silk Road". It also supplied abundant fish and salt products to the central plain regions of China. Today, Cangzhou is listed as an important port city in north China due to its Huanghua Port. However, at that time, the living environment in the southern part (close to the sea) of Cangzhou City was so poor that it was even worse than that

The Gate Tower of the Shanhaiguan Great Wall
(The black-and-white photo that I held in my hand was taken with my father in 1977.)

The Restored Map of the Ancient Buildings Hanging on the Gate Tower of the Shanhaiguan Great Wall

The Tablet of "The First Pass of China" Hanging on the Gate Tower of the Shanhaiguan Great Wall

The Rampart of the Shanhaiguan Great Wall

(The above photos were taken by the author in 2017.)

of Suizhong County; lacking suitable conditions for human habitat. I remember that the entire coastline was full of mud and had no sand, making it impossible to construct houses directly on it. Additionally, Cangzhou City had a lot of saline land containing high salt content, which caused low crop yields or even prevented crops from growing altogether. Since houses could not be built directly on this land, our job was to raise more than one meter by adding normal soil onto these saline lands measuring over two-thousand-plus square meters before building brick houses on top; ensuring they would not be flooded by rainwater during storms in most cases. The sands we used for house construction cost over one cent per pound which was even more expensive compared with millet prices at the time.

At that time, water resources were very scarce because the underground water was undrinkable due to its high salt and alkali content. Local people had to rely on the nearby artificial river to transport water from a long distance by boats. However, when it rained heavily, the river transportation would be affected, so people had to store water carefully. Therefore, water became so precious that local people normally had to avoid washing their children's faces or hands. What did the villagers eat then? Vegetables were so rare, and cabbage was almost the only cherished choice. After the autumn harvest, cabbages were simply cut into small pieces (either sliced into blocks like eggplant cubes or larger diced ones) without cleaning and dried slightly to remove moisture before being stored in jars due to their tendency of easy damage in air. However, this dish wasn't eaten often as vegetables were precious; instead, fish was mainly consumed by the villagers since it could be easily caught at almost no significant cost.

I stayed in Cangzhou City for one year in the early 1980s, and was then transferred back to my hometown—Qingdao City, where I have worked and lived for quite a long time since then. It's worth mentioning that in the early 1970s, my father had worked in Beijing for five years, hence I lived there with him for some time. My impression of Beijing is that there were many poplar trees, and the sound of the wind blowing through the leaves was beautiful. Besides that, not many other kinds of trees were planted. At that time, Beijing did not have any subway lines; however, it had numerous bus routes available although the buses themselves were relatively outdated. The prices of goods were stable in Beijing, and most daily necessities were readily available; additionally, dining out at restaurants was quite affordable.

There were numerous historical buildings; however, the residential structures primarily consisted of old houses which evoked a sense of oppression and antiquity. In contrast, the new houses boasted modern amenities such as indoor toilets and tap water; some even had TVs installed. I was astonished when I first saw a TV in Beijing—how could there be people inside a box? After the 1970s, washing machines made their ways to Beijing. It was imaginable at that time for clothes to be washed by machines; it seemed inconceivable for people living in places like Qinhuangdao City, Suizhong County, and Cangzhou City where they had never heard of such a thing before. I still vividly remember that in 1979 when managers from my workplace traveled to Beijing to purchase a washing machine, all my colleagues went into frenzy upon its arrival—just like today if anyone owns a robot at home-now that would truly make headlines.

Since then, I have had the opportunity to visit Beijing multiple times before eventually settling here permanently. One immediate impression that stands out is the sheer size of Beijing even compared to its past. This applies not only in terms of housing sizes or urban areas but also in terms of its vast scale and dimension of the entire city. Furthermore, compared to my previous experiences living in southern cities like Shanghai where my accent sometimes caused exclusivity, I have found Beijing to be remarkably inclusive. The northern accents are more widely accepted here compared to their southern counterparts. In Beijing, language barriers are non-existent, which has left a lasting positive impression on me.

Compared to the 1960s and 1970s, there wasn't much change in Qingdao City in the early 1980s. Due to their work requirement, my parents could only visit their hometown every four years, which gave me the chance to stay in Qingdao for a while. Overall, before 1981, the Qingdao City had a smaller area with many old buildings characterized by German and Japanese architectural styles from their colonization periods. By the sea, it was easy to dig for clams and oysters or catch fish; we could enjoy simple cooking after quickly cleaning them. However, it is no longer easy to obtain these seafood now; even if they are caught, it may not be safe to eat them. There were also many small crabs and sea anemones in the shallow sea; these "sea chrysanthemums" that bloomed all year round were incredibly beautiful. Nevertheless, these scenes are no longer visible.

The obvious changes began in the mid-1980s when Qingdao City was listed as

The Seaside of Qingdao City

(Photos were taken by the author in 2023.)

one of the first 14 coastal cities to open up to foreign investment. Consequently, the Qingdao Economic and Technological Development Zone was established in Huangdao District (now known as the West Coast New Area), where many foreign-invested enterprises built factories and office buildings, significantly driving up the standards and quality of construction. Additionally, a huge transformation occurred in the city center through demolition of many German and Japanese colonial-era buildings, while villages within the urban areas underwent renovation or new construction projects. Subsequently, the launch of the Qingdao-Huangdao Ferry served as a transportation hub connecting both the east and west coasts of Qingdao and linking with Qingdao Bay, further driving development along the Qingdao Bay Economic Belt which had since become an essential transportation tool for countless residents from both city center and Huangdao district.

Another turning point occurred in 2001 when Beijing won its bid to host the Olympics, with Qingdao being chosen as a co-host city for sailing events. Since then, economic development has accelerated in this city along with significant improvements in urban appearance and infrastructure; new residential buildings have continuously sprung up. The role of the Qingdao-Huangdao Ferry has gradually been replaced by those of the Jiaozhou Bay Tunnel and the Jiaozhou Bay Bridge, which have been successively built. My home is located in the old town

Historical Buildings—Street Scene near the Catholic Church in Qingdao City

(Photos were taken by the author in 2014 and 2023 respectively.)

area of Qingdao City where I can see many historical buildings during my daily activities. I have witnessed them undergoing multiple renovations. The revival of the century-old street has always been a topic for the long-time residents who lived nearby. Now, it seems that this street is reclaiming its former lively atmosphere. I am glad that some historical buildings have been preserved but also saddened bytheir loss of original appearance.

Historic District—Street Scene near Commercial Street of Zhongshan Avenue, Qingdao City
(Photos taken by the author in 2012 and 2023 respectively.)

I recently visited Suizhong County. After leaving Suizhong in 1970, I had never returned for 54 years. During my visit, I revisited all the places from my memory, except for one that was inaccessible due to being a military prohibited area. However, I couldn't find any trace of those places in my memory, not even the faintest resemblance. Perhaps I can only find some preserved photos at a photo studio or an archive center to revive my memories. All the urban planning and construction here is new.

The Suizhong Town Kindergarten was founded in 1951, named as Aiyu Park at that time (meaning "Love and Childcare Center"), which took in the Korean orphans who lost their parents in the Korean War. There were a total of four such war disaster relief centers in Liaoning Province, while Aiyu Park being one of them. In 1953, Suizhong Town Nursery was established based on Aiyu Park. In 1957, Suizhong County First Kindergarten was officially established. It was renamed Suizhong Town Kindergarten in 2021. The development of the kindergarten has been witnessed by three rows of small courtyards and two hundred-year-old ginkgo trees.

Suizhong Town Kindergarten—the Preschool I Attended

A Bustling Street Scene of the Most Prosperous District in Present-day Suizhong County (A)

A Bustling Street Scene of the Most Prosperous District in Present-day Suizhong County (B)

A Bustling Street Scene of the Most Prosperous District in Present-day Suizhong County (C)

Note: The above photos were taken by the author in 2024.

After years of hard work, we have achieved our current success. The changes between now and when we were young are indicative of the comprehensive progress of society. Taking Qingdao City as an example, urban construction is in full swing, and the streets and buildings are changing rapidly. The rapid development of the city has brought many conveniences to our daily lives, and there is continuous improvement in all aspects of clothing, food, housing, and transportation experiences. However, it is also regrettable that some historical buildings from our childhood or even earlier may not have been preserved or repaired due to incomplete policies or regulations. For many historical landmarks of this city that we remember, I often feel like they can no longer be found. When looking at old photos nowadays, young people may no longer recognize those familiar scenes.

Every city has its own unique features and special memories. An increasing number of people are now aware of the importance of preserving urban memory. Improving people's living environment is a long-term strategy, not an issue that can be addressed in 10 or 20 years; instead, it may require 50 or even 100 years to achieve significant results. If we could establish related regulations and standards for more precise and standardized guidance on constructing, repairing, and protecting historical buildings from a long-term perspective, I believe that every resident in every city will be able to find their own city memories, especially in high-density cities with high mobility in the future.

The Evolving Dynamics of Residences

by Li Pengfei

Before attending kindergarten, I lived with my grandparents in a two-story row house located in the Party Committee compound of Yongxing County, Chenzhou City, Hunan Province. These houses were allocated to government officials working for the local administration. Each household resided in a townhouse-like building with two stories: the ground floor consisted of two bedrooms, one living room, a kitchen and a bathroom; while the second floor had three bedrooms. Additionally, there was a large balcony and a small one. Behind the house, we could see the nearby cliffs. As my hometown was situated in hilly areas of southern China, the entire compound was constructed alongside mountains. Various residential houses for different departments of the local government were distributed at the foot of the mountains as well as on their slopes and tops. The landform resembled terraces and featured four levels of row houses with five households on each level.

The elderly residents would cultivate a small patch at the entrance yard to grow vegetables and also grew grapes on their rooftop balconies. Outside the residential area, there was a main road that branched into four ramps leading to row houses at different levels. The kindergarten I attended right at the foot of the mountain was very close to my grandparents' home; therefore, it was very convenient for me to go to school.

Both of my parents were teachers back then. When I was in primary school, I moved to live in the staff residential building where my mother worked. The primary school was situated in the heart of the county's territory. I remember it being called Red Flag Primary School, and now it has been renamed as Red Flag Experimental Primary School. During my time at this school, I experienced three changes of residence.

Front Facade of My Grandparents' Two-Story Row House

The Living Environment of Yongxing County, Chenzhou City, Hunan Province

At the beginning, there were only two buildings in the primary school: one for teaching activities, and the other behind was a staff dormitory. The teaching building was a two-story brick-concrete structure with walls made of adobe-like materials. The floors and railings on the second floor were all made of wood, creating an antique atmosphere. Before starting my second year at school, I lived with my parents in one staff quarter in the dormitory building. During those days, each family was allowed to occupy one room and shared a communal kitchen. Each staff quarter could barely accommodate five beds that were 1.5m wide in size. In my home, there were two beds against the walls, one for my parents and one for me, while the remaining space was reserved for other belongings.

After completing my second grade, a new four-story mixed-use building was constructed by the school. The former staff quarters were converted into supportive housing for schooling activities. I had the opportunity to witness the process of constructing the new building using precast cement. It boasted an interesting layout where teachers' apartments and classrooms were situated on each floor simultaneously. As you ascended the stairwell, classrooms could be found at one end while three teachers' apartments occupied space at the other end. These apartments featured a standard residential design comprising two bedrooms, one living room, one kitchen and one bathroom. The kitchen and bathroom were adjacent to each other, with access to the bathroom through the kitchen.

During that time period, buildings had no insulation layers or other insulation measures in place, nor were air conditioners installed. Unfortunately for my family residing in the westernmost position of this apartment complex, our unit was particularly hot in summer due to direct exposure to sunlight from the west, despite having a north-south transparency design (translator's note: the term "North-South Transparency Design" refers to a house that has windows in both the north and south rooms, providing optimal natural lighting and air flow) that facilitated air ventilation. Our commonly-used treatment was splashing water directly on the floors and then laying down a bamboo mat for sleeping.

In winter months, our heating solution involved burning briquettes with a ventilation duck that reached outside. Three blocks of honeycomb briquettes could provide warmth throughout the entire night. However, I remember experiencing carbon monoxide poisoning when I was in fifth grade. One morning, I woke up with a severe headache, which led my parents to suspect carbon monoxide poisoning. They quickly helped me leave home and stayed outside for half an hour while slowly giving me water.

I lived in this apartment from the second grade until I went to college in 2000, so most of my youth was spent here. What impressed me most was the extremely noisy environment due to students having classes on the same floor and running up and down the stairs especially loudly. While it was very convenient for my parents' work and my own schooling, it also meant a lack of freedom for children like me—we had to go home immediately after school. Additionally, parents could easily observe everything downstairs by standing on their balconies and simply calling us home. This kind of life had both advantages and disadvantages: there was almost no possibility for teenager rebellion, and we were unlikely to be late as it was so convenient to get to school. Furthermore, as long as we did not fall from the stairs, there were no safety risks at all. However, because of our lack of freedom, we children could only study and play on campus before going home directly after class. Most of the time my peers and I played at school, casually drawing with chalk on the ground or chasing each other for fun. The school also provided us with sports facilities such as parallel bars, horizontal bars, running tracks, sand pits, slides and the like.

Teachers' Apartments and Students' Classrooms on the Same Floor

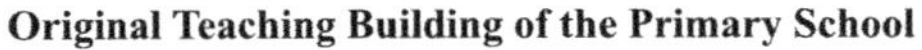

Original Teaching Building of the Primary School

Being Converted into A Residential Building

During this period of time, the primary school underwent restructuring and expansion; however, we continued to live here until I went off to college. My middle school years were spent at the County's No.1 Middle School, which happened to be located right next door to our primary school; hence there was no need for us to relocate elsewhere as it only took me five minutes on foot from home. When I was in middle school, I indeed longed for some personal space. I had a particular fondness for playing football, but at that time, my parents believed that it would hinder my studies, so they refused to let me play. This was a tragedy for me. Standing on the corridor of our apartment building, I could directly see students playing football on the playground of the No. 1 Middle School with just one glance. I remember watching my classmates indulging in playing football and sweating while all I could do was secretly kick the ball for 20 minutes before rushing home. It is unfortunate that I could not freely enjoy playing like other kids since my parents were much stricter about my studies during middle school.

The No.1 Middle School later expanded and acquired the campus of my primary school. The primary school relocated to a new location with newly constructed buildings and facilities, and my family also moved there. In the second half of 2000, after I left for college education, the new primary school completed its construction, creating a modern and standardized campus with trees and gardens. The school built a six-story residential building for teachers on campus, with each floor accommodating two households. This time, despite still being within the campus, the residential building was separated from the teaching buildings. We moved into a standard four-bedroom apartment on the new campus.

The New Campus of Red Flag Experimental Primary School

The Campus Scenery

The Residential Building on the New Campus of the Primary School

In the second half of 2000, I went to China Agricultural University for undergraduate studies and resided in a dormitory on the West Campus of the university. The men's dormitories accommodated six students in each room from various places such as Anhui Province, Hunan Province, Shanxi Province and Heilongjiang Province. There were two things that deeply impressed me about life in northern China. The first thing was experiencing a sandstorm for the first time. I remember one day while walking to class when suddenly I found myself stepping into a particularly yellowish world with small grains of sand getting into my mouth and nose; I couldn't even spit them out. At that time, I had no knowledge about sandstorms until my classmates informed me during lunchtime at the canteen. Sandstorms were quite severe back then: the sky turned yellow and there was sand everywhere, even inside my mouth. The second thing was experiencing freezing cold temperatures reaching minus 15 degrees Celsius during Beijing's winter months; even the water in the swimming pool would freeze completely.

After graduating from college in 2004, I rented a two-bedroom apartment in the Juyuan Community (translator's note: "Juyuan" literally means "Chrysanthemum Garden" in Chinese) next to my college and prepared for both the entrance exam for graduate school and job seeking. Initially, I shared the apartment with a college friend, but then he moved out; so my girlfriend and I lived alone. The happiest thing was that no one supervised me like before, and I could ride my bike through the college residential area to enjoy playing football every afternoon. It felt like a small recovery of freedom for my teenager years. The apartment of Juyuan Community featured a standard layout with two apartments located on each floor within a unit, enjoying north-south natural ventilation and lighting. The size of the apartment was about 65 square meters, and the rental price ranged from 1,300 to 1,500 yuan per month, which is an incomparable price compared to today.

Juyuan Community in Beijing

In 2008, the year Beijing hosted the Olympic Games, my girlfriend and I got married. Since the owner of the apartment in Juyuan Community wanted to take back the house for his own use, we decided to move to the "First City of Tomorrow" Community near the Lishui Bridge, where we rented an 80-square-meter two-bedroom apartment. This residential building had elevators installed. However, living there was somewhat noisy due to its proximity to a busy public road and poor acoustic insulation. We chose this apartment because it appeared quite new and clean with careful renovation and decoration. The owner was a doctor who initially planned on making it a wedding nest but changed his plan; he hardly ever lived in it after the renovation. At that time, I worked near the Asian Games Village in the North Fourth Ring Road. I commuted by BRT Line 3, which was convenient; sometimes I also drove myself to and from work.

In 2011, the owner of the "First City of Tomorrow" apartment planned to sell it, so we had to move again. This time, we chose to live in the residential community of

The "First City of Tomorrow" Community Near the Lishui Bridge

the East Campus of China Agricultural University. It was a brilliant four-bedroom apartment; however, the rental price was too expensive for us to afford, so we only rented the master bedroom with an attached bathroom. It cost us about 2,000 yuan a month, and we shared the living room, dining room and kitchen with other tenants. When we initially moved in, only the owner's elder brother lived there and later another family moved in; anyway, it did not feel crowded at all considering its large size. There was a underground garage in the community which made parking very convenient. Next to the Community stood a shopping center named Shengxi No. 8 where we rented a counter of about 80 square meters and ran a clothing store. Living here meant it was not only convenience for working at Asian Games Village but also for looking after our clothing store after work. However, my wife became pregnant after living here for about one year and we needed to rent a complete apartment. So we moved to the "New Dragon City" Community of Huilongguan area where we lived for another one year. Since we needed to rent an apartment alone then, rental price became a critical issue and it took us a much longer time to settle down in a two-bedroom apartment measuring 100 square meters at a monthly price of 3,900 yuan.

While renting, we also bought our own home in 2011. Located in the "Lingxiu Huigu" Community near the Zhuxinzhuang metro station of Line 8, the house was an 87-square-meter two-bedroom apartment with north-south natural ventilation

The Residential Community of the East Campus of China Agricultural University

and lighting, accommodating three households per floor. The price was about 15,000 yuan per square meters. At that time, since we had lost money running the clothing store, we had to borrow 30,000 yuan to pay for the public maintenance fund and property fees for purchasing this apartment; then, we further borrowed 30,000 yuan for simple interior decoration and furnishing. My mother-in-law hired a few decoration workers while I got involved myself. Most of the furniture were second-hand items from the Huilongguan flea market except for new mattresses, air conditioners and a refrigerator. My wife and I thought second-hand furniture was also very good because it retained little polluted smell while functioning well at super economic prices. Today, I am still satisfied with them:a coffee table at 100 yuan, a five-door wardrobe at about 300 yuan, a bed 500 yuan, a set of sofa 200 yuan, and even the integrated kitchen cabinet was second-hand - although its size was slightly unsuitable - we used some tools to adjust it to fit perfectly. Therefore, the total of 30,000 yuan successfully covered the expenses for decorating this apartment.

In 2013, we moved into this new apartment and our child was born. It was a cozy house to live in with wonderful amenities nearby. However, the traffic was very inconvenient, and I had to drive about 50 minutes from home to my work place.

Three years later, my company opened a branch office in Shenzhen City, and I was responsible for the development and management of the Shenzhen Office.

The "Lingxiu Huigu" Community

As a result, our family relocated to Shenzhen City in 2016 and has been living there ever since. In 2016, I purchased a small apartment near local schools with consideration for my child's future educational needs in Shenzhen City. Half a year later, I rented a four-bedroom suite measuring 150 square meters to accommodate our growing family size as my parents also moved from their hometown to live with us. Therefore, my wife, child, parents and I lived together. The monthly rental price was about 12,500 yuan. The main reason for choosing this suite was its convenience of commuting: the nearby Chegongmiao subway station served as a transfer station for four subway lines, making it convenient to travel between the east and west parts of the city. However, we later switched to a nearby two-bedroom suite measuring 80 square meters because my parents preferred to live separately. The rental price was about 8,000 yuan per month, which was much cheaper than before but didn't compromise our quality of life.

Looking back at these places where I have lived, my hometown is beautiful with lush mountains and clean rivers. However, it lacks convenience in terms of life and work conditions, making it more suitable for elderly parents to spend their remaining years at peace and comfort. When it comes to living in Beijing, my biggest concern is its commuting problem; daily long-distance commutes are a waste of time. Reflecting on all the residences I have lived in over the years, my favorite is the current two-bedroom apartment of 80 square meters in Shenzhen

City. It provides satisfactory convenience that perfectly meets my family's living demands. Although there are also disadvantages such as lacking of study space for myself or additional space for occasional stays by our parents, from the perspective of my own small family, I do prefer our current residence.

As for the housing itself, I believe it is attributed too many expectations as a fixed asset. In the past, because houses carried too many burdens, they would consume six "wallets" over three generations. I disagree with sacrificing many things just to afford a house. Fortunately, I have not purchased any property since buying that small apartment in 2016. If I had bought another one, it would have "smashed" me now with the pressure of repaying heavy loans every month while struggling to sell it for cash. "Houses are meant for living in, not speculating." If houses do not carry so much baggage and simply become places that shelter families from the wind and rain for peaceful rests, they would become much simpler and purer. Regardless, houses should serve people's lives; people should not become accessories to houses. Once freed from financial properties, houses can fulfill their pure function as shelters: providing necessary convenience for their occupants' lives such as easy access to schools for children. Today, I feel more and more people have started changing their perspective and eventually houses will return to their fundamental purpose as dwellings.

My Hometown, Embraced by the Blessings of Spring

by Li Xiran and Yang Jixia

Preface

A Journey From the "Hundred Herbals Garden" to the High-Rise Apartment;
Should the mountains drift,
Would the oceans change;
All lives rise and dance to greet the banquet of spring.

The "Hundred Herbals Garden" of My Childhood

"Oh, there you are, chasing chickens again..." I would always call out to my daughter from the door of our brick house, feeling worried about her aimless frolicking in the fields. As a child, she loved our expansive and level yard adorned with numerous trees, flowers and birds; it truly felt like a heaven for me who deeply loved nature.

In those days, my family lived in a brick house nestled in the countryside on the outskirts of the city. In my memories, the village had everything we ever needed: big ponds, wheat fields, orchards, dogs and cats, as well as many small shops. Children always enjoyed visiting those small shops to buy toys for amusement. Looking back now, it feels like that period of time was just yesterday. During my school years, I had to trek several kilometers before dawn every day to get to school. Back then, the road in front of my home was nothing more than an unpaved dirt path without any concrete or asphalt covering; It's unimaginable how difficult the road was during rainy days.

Our brick house was a bungalow with one spacious living room, three bedrooms, a kitchen, and a small but unique toilet that had a thatched roof; however, during that time, I greatly admired those who lived in two-story buildings. Our brick bungalow provided great comfort—it kept us warm in winter and cool in summer. The interior decoration and furniture were simple yet complete; however, they tended to accumulate dust easily which made them look perpetually dirty no matter how much cleaning we did. In my opinion, the countryside environment is preferable to the cities due to its pleasant scenery, fresh air, and charming old buildings. These elements establish my personal standards of comfort as I have high expectations for air quality. During those days, people were familiar with each other like family members because households resided closely together, and relatives enjoyed visiting each other on weekends—people maintained strong connections.

The rural market was located in the town rather than the village, and it had both a morning and an evening market. The elderly members in our family always liked to visit the market even before dawn, whether to buy or sell, and they would stay there until noon. Back then, people followed traditional living habits: early bedtime and early rising. The morning market typically started at dawn, offering vegetables and various items that people might need. During holidays, children loved going shopping at the market, picking up some of their favorite gadgets, and heading home happily to immerse themselves in their own small world.

A Spacious Home

Later on, we relocated from the countryside to the city for my daughter's primary school education because I believed it would be more convenient to live near her school, avoiding long commutes and early mornings. Our new home in the city was located in a six-story high-rise residential building surrounded by lush trees and well-designed gardens. The community also had a square with fitness equipment. It was quite different from our old bungalow in our hometown: instead of expansive courtyards, we then had narrow balconies; chickens, ducks, and geese had been replaced by birds flying across the rooftops. Our living space had significantly decreased. Furthermore, we also adjusted to going down multiple flights of stairs to reach ground level. Transportation became much easier but at an environmental cost as it was less eco-friendly there compared to before relocation. However, our home environment became clean and tidy with furniture being easily kept clean.

Despite having some items stacked up due to limited space, we were satisfied with its cleanliness and tidiness.

Finally, I was able to stroll along the smooth asphalt roads, and we owned both a car and an electric vehicle. The wide roads provided great convenience for transportation and shopping, significantly improving people's living standards. However, my child still missed her chickens and geese; after all, it was not practical to keep these poultry in an urban community. She longed for her childhood courtyard where tall fruit trees would often bear abundant sweet fruits. Despite the convenience of city life, we would travel back to my hometown on weekends to visit the elderly and relish the country life once again.

Over the years, from primary school to junior high school and then to high school, I have accompanied my daughter in witnessing the remarkable transformations and advancements of Wuhu City. With the demolition of old low buildings, numerous high-rise ones are erected; more and more residential buildings are now high-rise structures. Moreover, with the progress of science and technology, home appliances have become increasingly intelligent and advanced. Lying along both the Yangtze River and the Qingyijiang River, Wuhu city is renowned for its unique natural landscape characterized by a harmonious blend of "half mountains and half water". It is a well-deserved land of fish and rice; we enjoy the scenery of the "South of the Yangtze River", and its climate and terrain have been suitable for human habitation for thousands of years. Despite my personal naive wish, I do hope that our city will never be polluted but always maintain its healthy status.

Little Joys in Memory

When I was a child, the air was fresh and the countryside adorned with beautiful natural scenery. My greatest joy was running in the field at dusk and exploring adventures. In those days, there was no pollution: no developed industries or emerging industries; neither were there any heavy factory emissions. In short, the living environment was more beautiful and natural, fostering harmony between people and nature.

"Let the tides of the moon ease and swirl freely with the clouds; Mountains, rivers, forests, farmlands, lakes and grasslands compose a psalm of love on endless notes..." As portrayed in the song "One Earth, One Spring" (translator's note: it is the theme song for the 15th Meeting of the Conference of the Parties to the Convention on Biological Diversity (COP15)), humans and nature are interconnected elements of biodiversity and indispensable components within ecosystem. Just as President Xi remarked that "Lucid waters and lush mountains are invaluable assets.", we must strive for the long-term survival of our planet.

When my daughter was in primary school and junior high school, there were several years when the air quality was not ideal. Fog appeared in the areas south of the Yangtze River, and even haze occurred. People had to wear masks when going out in the morning. Although I couldn't pinpoint the cause of this air problem, I vividly remember how the sky lacked sunshine and appeared dim during daytime. The head teacher of my child always paid attention to the students' health status, especially during their commutes, and consistently urged them to wear masks after school. Nevertheless, these were only short episodes; overall, the air quality in south Anhui Province has always been quite pleasant.

Wuhu City, as one of the most livable cities in south Anhui Province, boasts an excellent geographical location with mountains and rivers, along with its long history. The Jiuzi birds (translator's note: Jiuzi is a kind of dove-like bird) and the Jiuzi Square are the most representative symbols for our city. I still remember when my daughter was small, the annual kite festival would be held in the Jiuzi Square, and an elder grandpa who lived downstairs would participate in it every year as a way to preserve traditional culture. The Jiuzi birds are considered lucky birds and serve as mascot of our city. Since my family moved when my daughter was in high school, we had not visited the Square for quite some time. When I finally revisited it, I noticed several changes: the original columns had been renovated, some trees were cut down, and even the fountain had been redesigned into a new style. Although it appeared neat and tidy, it no longer retained its original Mid-Century Modern (MCM) style.

The flowing time circles and the milky way rotates.
The sentience of all living creatures prospers and spreads like the wind.
The blue planet floats under the dome of the universe.
All lives rise and dance to greet the banquet of spring.

Social progress is accompanied by the development of science and technology, which brings changes to our time. However, the harmonious coexistence between humans and nature remains the fundamental principle of our universe.

Priceless Living Environment

Finally, the college entrance examination has been completed. Reflecting on the three years my daughter spent in a monotonous routine of commuting between school and home, she felt both familiar and distant while strolling through the city streets again. The light rail shuttles through this "City on Rivers", presenting itself as a captivating landscape.

Looking back, there used to be dust floating on the city roads. In contrast, at first glance now, you would notice a clean and orderly city with green belts, zebra crossings and isolation lines, creating a livable and pleasant environment. As time goes by, our living environment and transportation facilities have improved dramatically. The most eco-friendly change should be the adoption of electric vehicles. Additionally, the popularity of public transportation not only benefits the environment but also provides great convenience for citizens. As a result, traffic accidents have decreased, emissions from diesel and gasoline combustion have declined, and everything is progressing towards a better environment.

In the era of advocating ecological civilization, prioritizing the interdependence and mutual influence between human beings and the environment is crucial for constructing a harmonious ecological civilization. The sustainable development strategy serves as a necessary principle for building a community with a shared future for mankind, which also acts as a fundamental element for our survival on Earth within the universe. I deeply acknowledge the disparities in climate, landform and customs between the northern and southern regions; nevertheless, despite these differences, I believe that achieving harmony between man and nature and adapting to natural conditions are paramount in attaining genuine happiness for individuals, prosperity for nations, and revival for all nations on our planet.

The beautiful world, riot of countless colors.
One sky shared by different lands;
All roots derived from one origin.

Heaven and earth are one with me;
All beings on earth flourish together with me.
The sun and the moon thrive into eternity with me inseparably through.
The rosy clouds embrace me,
Amidst the most beautiful spring...

May our beloved hometown forever be embraced by the blessings of spring!

Transformation of My Home: A Living Story of a Post-80s Couple

by Ma Zhanrong and Han Ping

Our Old Hut in Xining

The city of Xining of Qinghai Province is considered the "Eastern Gateway" to the Qinghai-Tibet Plateau, boasting unique natural resources and colorful folk custom. My hometown is a small village in Handong Town, Huangzhong County of Xining City, while my husband's hometown is Xiazha village, which is not far away from my own.

Like all the old buildings in this part of the countryside, my home was a low hut with roofs supported by logs and walls made of straw and earth. It had only three rooms, but it had a large yard attached. Later, when red bricks became available, we used them to rebuild the front wall, making the entrance more presentable. During those days, we ate vegetables grown on our farmland, cultivated wheat and rapeseed in summer, and stored potatoes in the cellar for winter.

When I was young, no one had told me that I could have an opportunity to make a different life if I studied hard. Most girls around me grew up giving birth to children, doing farm work, and taking care of the elderly in the family. I thought everyone's life should be like that. My mother had a cheerful personality. My father couldn't read, but he insisted on sending his children to school and learning, even though he didn't know how to express the role of knowledge for our future. I dropped out of school when I was 15 since I couldn't keep up with my studies.

After getting married in 2005, my husband left home to work as a temporary laborer in gold mining for about seven months. His salary was extremely low, only seven or eight hundred yuan a month (100—110 USD). While he was away, I took

The Old Hut of Han Ping's Family

care of my father-in-law at home. We did farming and raised over 20 chickens. Occasionally, we would eat a chicken to have a better meal. When the family finally managed to save a little money, my father-in-law and I were able to renovate our hut. My father-in-law was very proud of it.

Life gradually got better. In 2007, I was pregnant for the first time. In 2008, my second child was born after the Beijing Olympics. Both of them were girls. We were able to settle down in our home with four small bungalows. I had hoped for a son after having daughters, which would have meant a perfect life to me. Anyway, I was satisfied with everything.

In 2011, I gave birth to my third child, a son. I was both excited and concerned because I worried that my father-in-law would favor the boy over the girls. Luckily, I was wrong. My father-in-law showered all his affection on my eldest daughter, saying that she should be the family's hope and attend college in the future.

Apartment in Xining

In 2011, in order to properly resettle the 16 villages of Handong Town and Dacai Town in Huangzhong County within the western area of Ganhe Industrial Park, the Administration Committee of the Park and Huangzhong County jointly invested in and constructed Kangchuan New Town in the nearby Chengdong Village of Doba

Town. The relocated households began to move successively since July 2012.

In August 2015, the Xining Municipal Party Committee and Government decided to utilize 6,540 mu (436 hectares) of industrial lands for the construction of Xining Horticulture Expo Park. The focus was on exploring a practical model for transforming "Lucid Waters and Lush Mountains" into "Mountains of Gold and Silver Treasures", promoting the construction of a model city characterized by green development and a happy Xining in the new era, and striving to explore a new approach to achieve integrated green development in ecologically fragile and underdeveloped areas.

In 2011, due to the construction plan of the industrial park, the place where we lived and several nearby villages were demolished and relocated. After that, my family was assigned an apartment in Kangchuan New Town. Later on, Xining Expo Park was constructed right on the location of my home village, and we visited it once and enjoyed its beautiful wetlands and a wealth of plants.

Photo taken in the Xining Expo Park

We did not move to the new apartment until our third child turned three years old, and we had been living there for five years. We led a comfortable life in Kangchuan New Town. The apartment was allocated by the government based on household size. Since we were a big family, a spacious 125 square meters apartment was assigned to us with one living room, one dining room, three bedrooms, and two bathrooms. The buildings in the community were six floors without elevators; however, we were lucky enough to win the lottery for a third-floor unit which made it much convenient for my father-in-law to walk up and down the stairs. All residents of the community belonged to the Hui ethnic group, and the conditions there were very nice—especially during cold winder days when heating was sufficient. When the heating system was turned on, our rooms became so warm that I could lie in bed comfortably after a day's work, being reluctant to get up at all. My fourth child used to have red cheeks in winter when we lived in Hainan Prefecture of Qinghai Province. However, once we returned to our apartment in Xining City, the child's skin would quickly recover within just two or three days.

Kangchuan New Town

Bayan Nur City, Inner Mongolia

In February 2018, my father-in-law passed away. My husband and I were very sad, but we could do nothing about it. There were so many children in the family who relied on us. I told my husband that we should stay strong and support the family through our own efforts. However, later on, we still couldn't make any money from our transportation business, so my husband and I had to move with our children to Dengkou County, Bayan Nur City of Inner Mongolia Autonomous Region to raise cattle.

Surrounding Areas of Kangchuang New Town

Residential Buildings Along the Street, Kangchuan New Town

In those days, my husband and his friends were raising cattle there. However, it was not easy to raise cattle due to the high cost of feedstuff. Besides, the cattle ate too much. We used up all our savings and we had to repay loans. There were almost no food at home, so I had to ask my mother for help with shame. Since then, I started working as a dishwasher in hot-pot restaurants, while my husband tried to earn

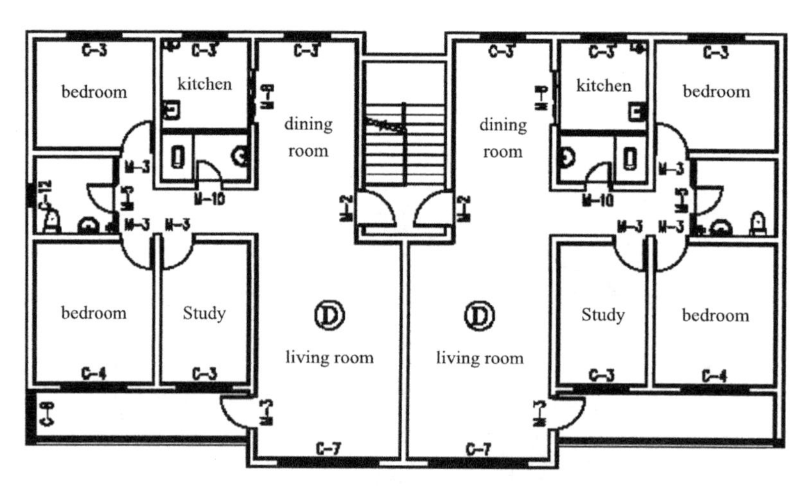

Apartment Plan (schematic), Kangchuan New Town

more money by taking care of other people's cattle along with ours. Despite these efforts, we could barely cover the food expenses for our children and the cattle.

In order to save money, we rented a shabby hovel in the County town for only 2,500 yuan (appr. 350 USD) per year. It was cheap, and the living conditions were very poor. One day, my children were playing mischievously and accidentally kicked the door, causing it to shatter into pieces and even breaking some bricks. The winter of Inner Mongolia was extremely cold with relentless winds that persisted even in February. our hovel had no heating system and was poorly sealed. So, even with a fire stove in the room, it made little difference as the hot air could hardly be retained. Consequently, when we returned to Xining City, we found its winter weather comparatively much warmer.

Although life was hard, people there had changed my mindset during the days I worked outside. I realized that they lived an active life: although they did not demand a high level of accommodation or lifestyle; they were very capable and diligent. They were also content with simple food. People there were particularly clear-minded and believed that young people should mature and develop their life skills before getting married. This instantly educated me, and I thought it was true. I also wanted my children not to rush into marriage like me.

One year later, the loan expired and the cattle became infected with infectious diseases. There was no choice but to burn the sick cattle and consider returning home. My husband blamed himself a useless man for leading such a life without earning anything in a year. I comforted him by saying that the five of us had been using the rancher's pasture for free for an entire year. Wouldn't it be wonderful? My husband laughed, and so did I.

Hainan Tibetan Autonomous Prefecture

We arrived home and felt that Qinghai was indeed a good place; however, life didn't allow us to rest. In 2019, my husband took the whole family to Hainan Prefecture in Qinghai Province, a place that might bring us hope. At that time, I was thirty-two years old and unexpectedly became pregnant again. It turned out to be a girl. The doctor said an abortion would be very dangerous, so I thought,"All right, let things happen then."

We took out a loan and opened a small restaurant in the town, and the business was doing well. The first and second floors were kept for customers, while our family of six lived on the third floor, which had a roof made of colored steel tiles. The rooms were sparsely furnished. A hallway was separated by boards, and the remaining space was divided into three rooms where we placed bunk beds as bedrooms. Although there was a toilet, it was non-functional, so we had to use nearby public toilets. We cooked in the kitchen on the first floor. The house had central heating

Outside the Restaurant of Gonghe County, Hainan Prefecture

installed, but it provided little warmth since its exterior walls were merely standard brick walls of a width of 370mm, simply painted and without insulated layer. Nevertheless, it was still much better than the hovel we lived in Inner Mongolia.

We needed to travel back to Kangchuan New Town for the New Year holidays. However, we were always fatigue from traveling by bus with four children, so we managed to buy a car. Although it is still a difficult time, I think I will feel happy for now when I get old. I used to feel bitter when we did farming years ago, but now I recall those days as good times. I remember my father-in-law and I could hardly have any meat in summer. One day, my husband came back with two hundred yuan (appr. 28 USD) and the two of us rode a motorcycle in heavy rain to buy a leg of lamb and prepared a meal of hot meat dumplings. How happy and wonderful it was! Anyway, now I can eat meat whenever I want.

My mother's house has also been renovated in recent years, featuring exquisitely carved beams and a bright sunlight room that make it beautiful and grand. The family can bask in the sunlight together or engage in physical exercises, making our lives happier every day. Each stage of life brings its own happiness, and contentment leads to long-lasting joy.

The Old House of Han Ping's Parents, after Renovation

My children are studying at the same school, and they are fortunate to enjoy nine years of compulsory education with waived tuition fees provided by the government. We are living in a good time, and this is our greatest blessing. Our responsibility is to guide the children towards becoming descent individuals; this would be a true measure of success for us. Despite not having much free time available, our family occasionally takes trips to the mountains for fun, which brings us immense joy. The utmost thing for a family is ensuring everyone's safety, health and happiness.

A Vision for the Future

With gradual improvement in living conditions, we would love to take our family to see the outside world in the future. The first stop should be Tiananmen Square in Beijing. If the restaurant business is doing well, we hope to buy a small suite near the restaurant so that the children can live in warmth and comfort, and we can take warm baths in the evening. Once all loans are paid off and our children have grown up, our daughters may have opportunities to attend college and secure good jobs instead of getting married early. I do not anticipate excessive fortune; I simply want my children to appreciate everything achieved through hard work after overcoming all the difficulties of these years.

* The exchange ratio between USD and RMB Yuan in this book is 7.2.

Memory of My Hometown

by Niu Cunqi

My hometown is located in the countryside of northern Henan Province, where the village and farmlands spread along the previous drainage basin of the Yellow River. Dunes stretch from west to east in its north, and jujube trees and other drought-resistant fruit trees are commonly seen on these dunes. The cultivated lands consist of typical sandy fields with poor water and fertilizer retention ability. In the south of the village, arable lands are half sticky and half sandy, with Jindi Embankment and Jindi River further to the south. Agricultural irrigation heavily relies on machinery wells.

Wells for Living

In my childhood memory, there was a dry well for drinking water to the left of my home. A dry well had to be dug into the ground until it reached the underground water layer. Then, it was paved with stones and sand before building brick rings extending up to the ground level, resulting in a total depth of approximately ten meters. To ensure sufficient water supply, the bottom area of the well needed to be relatively large with a diameter of over ten feet and gradually narrowed towards the upper wall until it reached a remaining diameter at the wellhead of about two meters. This created a shape resembling that of a bobbin—small mouth but large belly. The wellhead was constructed with green stone stripes in a square shape that exactly resembled the Chinese Character "井" (Well).

Water was fetched from the well using a twisted rope made of willow bark, with a thickness similar to that of a duck egg. A hook made from a tree fork was attached to the end of the rope, which was then hooked onto a tin bucket and lowered into the well. By swinging or weighing over the bucket, we were able to fill it with water and pull it back up. However, swinging or weighing over the bucket required technical skills as there was always a risk of dropping the bucket directly into the bottom of the well or not being able to retrieve any water at all. Over time, deep grooves have been worn down on the green strip stones by the well rope, leaving visible marks of years' friction.

The well water was brought home by a carrying pole with two hook ropes at each end. At that time, I was still too young to carry much water. In fact, I could only lift the bucket off the ground; as a result, I had to wrap the hook ropes around the carrying pole to make them shorter. Furthermore, my strength was barely sufficient to pick up two and a half barrels.

In drought years, water could hardly be fetched from the dry well, and villagers had to rely on a mechanical well located in the middle of the village. The diameter of this well was much smaller, and a pulley was installed on it, which indeed saved labor. Compared to the dry well, it required more skilled technique: firstly, the well rope could not be swung but only repeatedly weighed up and down to fill up the bucket; secondly, when water was brought up to the mouth of the well, one needed to handle both the pulley with one hand and pull the well rope with the other hand in order to align them properly, which was difficult for children to master. Furthermore, its location was over half a mile away from my home, so I dared not go there alone without someone helping me draw water from that well.

A third well was located further east in the village right outside of the village school. However, due to its bitter and salty taste, water from this well was undrinkable; instead, it was used for mixing mud for construction or cleaning the dusty streets.

Later, a rubber wheel trolley was invented. I should be credited as the first person to create a shelf for fetching water. It featured a support frame made from four connected logs attached to an axle, with a long wooden bar above it for hanging buckets on both ends. This invention allowed two pairs of buckets be pulled at a time, saving labor and improving efficiency. It remains the most impressive invention in my memory. Over time, water consumption has become more convenient. Trolleys were then used to transport gasoline barrels for water storage; pressure wells were dug in villagers' own yards; a water tower was built near the machinery well; tap water was connected to households; and water pipes were

installed in kitchens. Solar panels had been installed for bathing purposes.
However, with improved living conditions came increased consumption of not only drinking but also cleaning in kitchens and doing laundry with washing machines. As a result, the amount of waste water had significantly increased, and proper drainage had become a prominent issue to address. Various types of wastewater lacked appropriate arrangements for discharge, and domestic sewage often flooded along the alleys and streets in the village.

Saving Money to Build Our Own House

Our village was originally called Gaozhai Village (Gaozhai means "stockade" or "fort"), and it was renamed Gao Village during the campaign of "Learning from Tachai in Agriculture" in the 1960s. There were stockades encircling the village, but I don't recall there being a fort gate. My family lived at the west end of the village, which was also the first farmhouse closest to the edge of the stockades. A drainage ditch ran right outside of the stockades. Usually, there was no water in it, but during heavy autumn rains, water from the northern field would flow into the ditch and further down to Jinti River in south.
When I could remember, four generations of my family resided in a narrow courtyard with three eastern rooms, two northern rooms and two western rooms. In reality, there were not many separate rooms; instead, they were simply divided areas by beams or partitions. Therefore, each room was very small and would only be furnished with a bed and a small table. Beds were either earth *Kang* (*translator's note: traditional brick/earth bed that can be heated for winter in northern China*) or a few wooden boards, while tables were made of earth platforms with stone plates on top. With low ceilings and windows covered with oiled paper, the rooms were always dark and damp.
The three eastern rooms were the best among all. Half a meter of the lower part of the walls were made of gray bricks, above which adobes wrapped with bricks formed the upper part. The north adobe wall served as a partition for a separate single room, while the south partition consisted of single-layer brick and wooden cross grids with a single door leading to the inner room. The layout of "one illuminated and two dark" characterized these three rooms. Lime-seamed bricks and purple-red painted grilles added an elegant touch. My parents renovated these

eastern rooms on the original site of the old rented house when I was five years old. The two northern rooms were made of adobe, and a stove was located right at the door of the room on the eastern side. My great-grandfather lived in the room on the western side, which was his ancestral property. The western house was an earth-mound hut that had been divided into two rooms by half a gable and half a beam structure. The inner room was installed with a large earth *Kang* where my grandmother resided with my eldest younger sister. There was only a small window present. The western hut had been stacked up by my parents with earth collected from the stockades—one shovel after another and mixed with water.

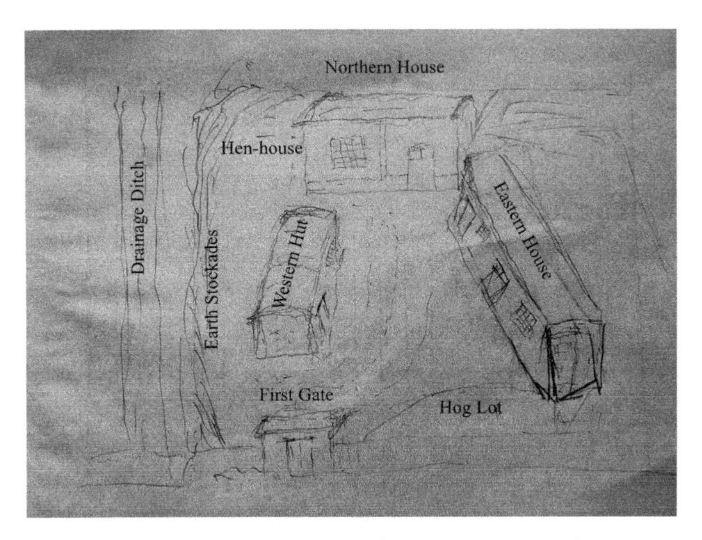

Original Courtyard Layout (Sketch by the Author)

When I was about ten years old, I first remembered the scenes of building houses at home: while our western hut was being demolished, the northern house was simultaneously being rebuilt.

What impressed me most was the laying of the foundation. A stone roller, used for grinding grains during harvest seasons, was erected and tied up with iron wires to form an "井" -shaped (intersection parallels) bar. Eight strong men carried it as a huge hammer to tamp down on the foundation. Half a meter of flat bricks were spread on the leveled ground with brick slag filling in the blank, above which bricks wrapped around inside adobes. At a height of ten feet, ridges were constructed on herringbone-shaped beams and covered with a red tiled roof. It took only one day to lay the beams and two days to install tiles; thus, a red brick and red tiled house was erected in just a few days. During that time, after school, I would climb onto shelves to pick up bricks thrown up by adults, and my little sisters could also help adults smash brick slag and fill in gaps.

That year, our family finally had a house with three rooms that had tiled roofs and red brick floors. The excitement was no less than that of today's rooms decorated with ceramic tiles and wooden floors.

During the early days of reform and opening up, the village planned new streets and built new brick kilns. During the New Countryside Construction Campaign, our family was allocated a new plot for a homestead; therefore, we built two additional courtyards. In fact, our hardworking parents have been saving money for over 20 years to build houses for their children.

The Original Courtyard (Photo taken in 1999)

The Old House in the Original Courtyard (Photo taken in 2004)

Screen Wall (Built in 1992. Photo taken in 2007)

My younger brother has fully inherited our parents' enthusiasm for residential houses. With the improvement of economic conditions, he renovated the old house left by our parents in 2011, and further saved money to buy and decorate apartments in the county town for future generations, firmly inheriting the traditional attachment Chinese people have towards residential houses. The current courtyard of my hometown was renovated in 2011 on the original site of the old house characterized by "five new tiled rooms with a corridor" that our family had resided in for 30 years. Its main structure is one and a half stories, with both upper and lower floors made of cast-in-place concrete for ring beams and ceilings, which emphasizes structural strength. Upon entering the main hall, which is nearly 50 square meters in size, there are two living rooms on both sides with different functions. A chandelier hangs from the center of the ceiling while spotlights are embedded in suspended ceilings around it. The four walls are coated with wall paint, and ceramic tiles cover the floors. Large glass windows with window casings and the interior compound doors with door casings significantly improve both functionality and aesthetics of the living environment.

After graduating from college, I moved to the city for work. At first, I lived in a dormitory provided by my work unit (translator's note: "work unit" refers to an employer, particularly state-owned companies or organizations in mainland

The Current House (Built in 2011. Photo taken in 2015)

The Current Screen Wall (Built in 2011. Photo taken in 2023)

China). After getting married and having children, my family resided in various apartments with public corridors that were also allocated by the employer. Initially, we used briquettes within the corridor area, but later switched to a natural-gas stove for cooking in a communal kitchen. Throughout the past 30 years, our housing conditions have gradually improved through stages such as housing distribution programs, purchasing "Reformation Housing" properties at employees' discounts, and buying commercial housing directly from our employer. Our living environment has also undergone earth-shaking changes.

The Current Courtyard (Built in 2011. Photo taken in 2023)

From a historical perspective, the continuous pursuit of infrastructure upgrades is deeply rooted in the mindset of Chinese people. It's no wonder we have earned the nickname "Infrastructure Giants". The construction of new cities, districts, roads, and bridges, all reflect our nation's aspiration for a better life from various specific perspectives.

Developing Tourism for Poverty Alleviation

—A Story of Dazhai Village, Longji Town

by Pan Baoyu

Situated in the east of Longsheng County Multinational Autonomous County (hereinafter referred to as Longsheng County), Dazhai Village is located in the core spots of the famous Scenic Area of Longji Rice Terraces, Longji Town, Guilin City. The Village is 36 kilometers away from the County town and 23 kilometers away from the township government. It consists of 15 villagers' groups, with a total of 293 households and a population of 1,227 people. Over 98% of the population belongs to the Yao ethnic group. The village cultivates an area of 454 mu (*translator's note: 1 mu equals to 666.7 square meters, and 454 mu is approximatly 30 hectares*) of irrigated farmlands and 886 mu (*appr. 60 hectares*) of paddy fields. For generations, the ancient Yao people have cultivated on the hillsides, creating a great group of terraces with layers on top of each other, viewed just like the Jacob's ladder.

From Working Outside to Starting a Business Back Home

About 20 years ago, Dazhai Village was terribly poor. In 1994, Mr. Chen Xiaoqing, a famous TV director of CCTV, visited us and made a documentary film that truthfully recorded the impoverished and underdeveloped scenes in the mountain villages of Yao people, as well as the lack of educational opportunities for children. "Half the iron pot, half the house, half the bed and half the nest" was the true portrayal at that time.

In 1998, although my family was very poor, it was not among the poorest families in the village. The Village Ccommittee once distributed donated clothes to needy families, and in order to receive a pair of pants, I had to explain to the committee

The Documentary Film of CCTV: *Longji Town*

members that I needed them for work in Beijing. However, the pants were so long and big that I had to shorten the leg openings to wear them.

In 1999, I came to Beijing to work as a carpenter in the Chinese Ethnic Culture Park, where we constructed all the wooden structures. At that time, I earned 12 yuan (appr. 1.7 USD) a day, and almost all the money I received was spent on visiting scenic spots. Amid the crowded tourists, I climbed the Great Wall during the National Day and felt deeply inspired by the remarkable feat of human engineering. The Great Wall is man-made, just like the terraces in my hometown. I felt the amount of labor invested in constructing terraces is comparable to that of the Great Wall. I dreamed that one day there would be as many visitors exploring the beautiful terraces in my hometown as those visiting the Great Wall.

Picture Taken at Tienanmen Square When Working in Beijing in 1999

After spending a year in Beijing, I returned home with a broadened vision when hearing that my hometown was planning to develop terrace sightseeing. Upon arriving home, I looked up and down at the small huts where the villagers and I resided, and realized that Dazhai Village would have no prospects if it was confined to the mountains. At that time, there were no driveways and villagers had to rely on manual labor and animals to get in and out of the mountains. However, the spectacular scenery of the terraces had already attracted some photographers, leading me to speculate that the homestay business might become popular in the future.

Therefore, around 2000, my family and I built the first homestay hotel in the village. We could barely afford the cheapest tiles and had to produce them ourselves. I borrowed 3,000 yuan (appr. 417 USD) from the local rural credit cooperative, sold our cattle, and finally managed to gather more than 10,000 yuan (appr. 1,389 USD). It took us three years to hand-build the hotel. On June 26th, 2003, the Dazhai Village Tour officially opened to visitors. On that very day, my homestay hotel was also opened for business, charging 15 yuan (appr. 2.1 USD) per room per day. The villagers realized that they could earn money by running homestays so they also started constructing their hotels. In this way, tourism in our village gradually developed.

In those days, mobile phones could not take pictures and there was no Internet. Shutterbugs and hikers had cameras, but how to attract them to visit our village and take pictures? I thought of writing a pair of couplets to post, encouraging the tourists to take photos. I composed a couplet that read: "Gratitude for the Improved Utilities and Infrastructure Achieved Through Government Efforts; Commemoration to Ancestors for Creating Spectacular Terrace Scenery", and affixed it onto the gate. When tourists arrived and saw the couplet, they found it novel and proceeded to take photos one by one. As a result, photography contributed to creating a unique experience among the tourists.

Explore the Way for Poverty Alleviation through Tourism

Cooperation for Terrace Tourism Development

In 2003, Dazhai Village initiated a cooperation with Guilin Longji Tourism Development Co., Ltd. (hereinafter referred to as "Longji Company"). Longji Company and Dazhai Village signed a three-year agreement: Dazhai Village took the terrace landscape as resources shares; while Longji Company should be in charge of branding, management and operation of the terrace landscape in a consolidated manner. According to the agreement, the village could receive an annual maintenance service charge of 25,000 yuan (appr. 3,472 USD), and villagers were required to grow rice and maintain the terrace landscapes.

However, Dazhai Village faced a dilemma: although the tourist road had been officially opened to traffic, Dazhai Village needed to repay the 118,000 yuan (appr. 16,400 USD) of debt for constructing the road. Our village could not afford such a large amount of money. Some village committee members proposed to reach out to a rich boss to pay off the debt at a lump-sum settlement, while the maintenance service charge collected from the terrace tourism could be refunded to the boss in 10 years.

As a normal village committee member at that time, I rejected the plan with the consideration that villagers, rather than the business owner, should be the primary beneficiary from the terraces since they had been planting the fields with hard work over the years. Besides, an increase in tourists would benefit us with potential elevation in maintenance service charge in the coming years. So, I put forward another plan: Longji Company paid a total of 75,000 yuan (appr. 10,380 USD) in advance for three years' maintenance service. Eventually, my proposal gained support from the villagers. Later, I approached the Chairman of Longji Company and the township government to propose this plan, and Longji Company agreed to lend money to Dazhai Village, helping us to pay off the debt. In this case, my proposal not only benefited villagers by sharing the tourism revenue but also enhanced the cooperation with Longji Company. Therefore, the villagers elected me as the Chairman of the Village Committee by high votes during the village election in 2005.

When I first became the Chairman, the Village Committee was still burdened with debts. Dazhai Village was labeled as a national-level impoverished village without a penny of income for the collective economy, while the villagers still relied on outside support to make a living. After the implementation of "Village-Based Poverty Alleviation" campaign in 2005, Dazhai Village gradually constructed roads and bridges, along with public cultural complex buildings and parking lots. In 2006, Dazhai Village was selected as a pilot for the development of "Village-based Poverty Alleviation" and "Socialist New Countryside", which received strong support from the poverty alleviation departments of both the Autonomous Region and Guilin City. As a result, significant improvements were achieved in infrastructure construction.

The Drying Clothes Festival

In 2005, terrace tourism was developing but still not well known. I figured how to spread our Red Yao culture to more people (*translators' note: Red Yao is a branch of Yao Ethnic*). Then, I thought about a traditional Yao festival on every June 6th of the lunar calendar, which is as important as the Spring Festival. On that day, every family would have their daughters and sons-in-law back home, and the elder people would sun their traditional costumes, calling the "Drying Clothes Festival". Actually, there are various celebration activities such as weaving and spinning on June 6th for Yao people, which tourists could appreciate. So, in 2006, I planned to warm up the atmosphere for the festival, and organized the villagers to come together and held the Drying Clothes Festival in a grand way.

Due to the lack of funds, I invited Longji Company to join us. At the beginning, the company's chairman planned to sponsor us with 1,000 yuan (appr. 140 USD), since he thought the festival was merely about drinking and feasting together. However, it was not my intention—what I truly expected was to cheer up the festival atmosphere and making it a grand celebration; while managing with 1,000 yuan (appr. 140 USD) would be quite challenging.

In order to make the Drying Clothes Festival a success, I visited Guilin City in 2006 to meet with the Chairman of the Photographers Association and the Chairman of the Art Photographers Association, and discussed with them about the festival and their potential participation. They agreed to invite some photographers to our village for shooting and publicity.

At that time, there were 46 households in our village running homestay businesses. The Village Committee suggested raising 200 yuan (appr. 28 USD) from each of them as sponsorship. I promised that their business would be fully booked on the night of June 6th, and if there were any vacancies, the Village Committee would refund the sponsorship. Finally, the suggestion won support from the villagers. The first "June 6th Drying Clothes Festival" of 2006 proved to be a great event and continued its success in the following years. The Festival achieved a significant scale and brand recognition until it was suspended for a year due to the Covid-19 pandemic in 2020. Dazhai Village has become a popular scenic spot with a rapidly rise in fame and a booming number of tourists, partially owing to the success of the Drying Clothes Festival in the past ten years.

The Drying Clothes Festival of Dazhai Village, Longji Town

Various Activities during the Clothing Drying Festival

Joint Efforts and Sharing of Tourism Benefits

Dividend Management

There were also some management problems with the increase in tourists. For instance, to attract customers, some villagers would take tourists on alternative routes to evade entrance tickets, which resulted in chaos in the management. To address this issue, the Village Committee set up an assistant team to collaborate with Longji Company in apprehending ticket evaders and requiring them to compensate losses. This led to complaints from some villagers that the village was helping the Company in making money instead of benefiting the villagers.

The original dividend plan between the Village Committee and Longji Company from 2004 to 2007 was that the Company should pay the Village Committee 25,000 yuan (appr. 3,500 USD) per year from tourism revenue as dividends. However, the villagers showed little interest in it, leading to a state of management chaos. So, in 2007, through the Representative Committee of our village, I proposed a new scheme for profits sharing, which was supported by the villagers. When renewing the cooperation agreement, we brought up an adjustment plan with the Chairman of Longji Company based on the concerns that sticking to the previous dividend scheme would only perpetuate the management chaos. Finally, the Company agreed to pay dividends according to a certain proportion of ticket income. It would pay 150,000 yuan (appr. 20,760 USD) as guaranteed dividend; and if ticket income exceeded this amount, 10% of the extra income should be further shared with Dazhai Village, with 7% from core scenic spots, and 3% from non-core scenic spots. In 2008, the Village Committee received 147,000 Yuan (appr. 20,345 USD) from ticket income and the Company compensated for the inadequate sum to 150,000 Yuan (appr. 21,760 USD) according to the new agreement. In 2009, the dividend increased to 156,000 Yuan (appr. 21,590 USD) and it boosted to 7.2 million Yuan (appr. 1 million USD) in 2019.

Dividend Distribution Ceremony of Dazhai Village in 2016

(An Achievement of the "Tourism Relieving Poverty Campaign")

Terraces maintenance

The primary work of the entire village is to cultivate the terraces, and farming is a kind of scenery maintenance to preserve our landscape resources. However, with more and more villagers venturing into tourism business, some of them have gradually abandoned terrace farming and maintenance.

Therefore, the Village Committee set up a terrace maintenance team in 2008 to encourage villagers returning to farming. Some families have indeed faced difficulties since they only had elderly members and children in the family, without any additional labor force available,which made it impossible for them to engage in farming. In such cases, the terrace maintenance team helped them cultivate and maintain the terraces.

Furthermore, Longji Company signed terraces maintenance contracts with all relevant villages of the scenic area, and put forward an award and compensation plan. For example, Dazhai Village should be responsible for terrace cultivation and landscape protection, and if over 360,000 tourists visited the scenic area, Longji Company would reward the villagers who cultivated terraces with 100 yuan per *mu* (*translator's note: approximately 200 USD per hectare*). Considering sustainable tourism development, Dazhai Village had worked out a distribution plan to allocate the revenue paid by Longji Company for terrace maintenance service: 50% was rewarded to the villagers who cultivated the terraces, 20% was distributed based on population, another 20% was distributed equally among households, 5% served as

Traditional Farming Cultural Activity of "Plough Side by Side" for Spring Tillage

compensation for farmers whose mountain forests were occupied or damaged by road construction, and the remaining 5% was allocated to the village's collective economy. In 2013, the distribution scheme was re-adjusted and increased to subsidize farming by 70%, distributed funds on household by 12%, on population by 12%, 3% to farmers whose mountain forests were occupied or damaged by road construction, and the remaining 3% to the village's collective economy. According to this plan, each *mu* of the terraces received 8,900 Yuan in 2019 (*translator's note: approximately 20,000 USD per hectare*).

With increasing compensation on terrace resources, more and more villagers actively participate in farming activities, and there is no longer a need for the terrace maintenance team. Each year, Longji Company would measure the areas of paddy fields and pay bonuses accordingly.

Turning Dazhai Village's Golden Pit into A Mountain of Gold

Construction of the Cable Car

The income of Dazhai Village increased significantly since the construction of a cable car project in 2013, along with a noticeable increase in tourists. The cable car company originally planned to construct it from Erlong Bridge to Ping'an Zhuang Ethnic Village. However, Ping'an Township Government was reluctant to carry out the plan due to the concerns about potential impacts on their lands. Therefore, they did not welcome the cable car company. I once had a tour to Zhangjiajie (*translator's note: a World Natural Heritage of China*) and witnessed a great number of tourists there. When I came back, I reflected on why we could not attract so many tourists here? A possible reason might be the inferior traffic conditions. I paid a visit to the County government and had a discussion with the Deputy County Mayor regarding the construction of a cable car project in Dazhai Village. The villagers did not agree at the beginning, especially those who made money by carrying visitors' sedan backpacks, being afraid of robbing their business. When the construction team arrived, they were even driven out by the villagers and hindered from accessing the construction site for over 4 months.

Afterward, I discussed with both the Village Committee and the Village Party Branch Committee for this issue. I believed that cable car riding would be a convenient means of transportation for tourists to ascend and descend the mountains easily, without depriving the villagers of their income. The best place to enjoy the terrace beauty should be from a high position. Carrying sedan backpacks on the steep mountain paths was hard work and dangerous, earning little money. Furthermore, upon the completion of the project, there would undoubtedly be an increase in visitors, even allowing children and elderly people who were previously unable to make it up to the mountains.

After talking with the committee members, we persuaded the 31 representatives of the villagers one by one. Finally, 25 of them signed the agreement with the cable car company, and the construction got on the right track. The project was accomplished for trial operation at the end of 2012, and villagers' income has been doubling every year since 2013.

The cable car company allocates 7% of its revenue to Dazhai Village and the County government, with 40% going to the County and 60% to the Village. Besides, villagers are entitled to free rides. When negotiating with the cable car company, I introduced the concept of "resource charge", emphasizing the contribution of terrace landscapes, which helped us secure the dividend from the cable car company.

(Translator's note: The mountain area where Dazhai Village is located was known as the "Gold Pit" because gold was discovered and extracted in the past. However,

Golden Pit Cable Car for Dazhai Terrace Sightseeing

due to various reasons, the villagers did not benefit from the gold mine and remained impoverished. The successful operation of tour business has significantly increased revenue for the villagers. The picturesque scenery of "lucid waters and lush mountains" is an invaluable asset, turning the place into a true mountain of gold.)

to Create Another Golden Week

Paddy field harvesting happened after the National Holiday (*translator's note: also known as "Golden Week" for tourism business*) every year. In 2017, considering that many tourists would still pay a visit after Golden Week and villagers had heavily invested in constructing homestay hotels, requiring income growth from tourists, we came up with the idea of postponing the rice harvest schedule, resulting in the creation of another Golden Week.

I went to solicit opinions from the villagers, and over 80% of them agreed. The Village Committee made a commitment to Longji Company, publicly announcing that tourists could still enjoy the scenery of the golden terraces even after the Golden Week. So, the rice was retained in the fields until October 16th in 2017, while the dates were delayed to October 20th in 2018 and 26th in 2019 respectively. But in any case, food had never been wasted merely for its scenic value. Now there is a new plan: the villagers would harvest only the ears of rice and leave the straw in the fields to maintain the golden landscape for an additional month or two.

Golden Terrace Landscape

Rice Harvesting in the Terraces

Future Outlook

Dazhai Village has been committed to the development of tourism for poverty alleviation for more than 20 years, witnessing earth-shaking changes in terms of roads and other infrastructure, sightseeing convenience, accommodation capacity, and villagers' awareness and knowledge. Now everyone understands that the terrace landscapes are their "Iron Rice Bowl" (*translator's note: a Chinese contemporary metaphor that refers to a lifelong secured income resource*), and the villagers' awareness of protecting the terrace landscapes has also been promoted.

Tourism is developing well in our village. The agreements with Longji Company and the cable car company secure tourism dividends and farming subsidies for the villagers. Additionally, villagers actively engage in agritainment, catering and accommodation business. In fact, 80% of them have been involved in the tourism service business, experiencing a steadily increase in income and bringing significant changes to every household. Initially being poor with almost nothing, we now live a comfortable life with refrigerators, washing machines, color TVs, and air conditioners. When smartphones appeared, almost every villager bought one. Life is improving day by day. I am also very happy that my dream has gradually come true. A couplet originating from my heart goes as follows: The Beauty of Terraces is Shared by Friends Worldwide; The Eco-Friendly Tourism Benefits Everyone. Horizontal scroll: Great Happiness for All.

A View of the Dazhai Village, Longji County

For the future development of Dazhai Village, the primary issue is to continue protecting and preserving Yao culture. We also encourage villagers to inherit folk culture through a dividend distribution plan. For instance, men, women and children will receive a 4% reward if they wear the Yao ethnic costumes throughout the year.

Regarding cultural inheritance, we have planned to build a history museum for the village. The Village History Hall will exhibit the initial working tools and antiques of our ancestors, as well as old photos that bear witness to Dazhai Village's transition from an extremely poor village to a rural resort. It will pass on these stories to the future generations. Another reason for constructing the Village History Hall is that not only surrounding counties but also remote ones from other provinces and even abroad visit us to exchange ideas and learn about how to develop rural ecological tourism. These exchanges also serves as demonstrations of mutually beneficial cooperation between enterprises and villages in preserving Dazhai Village's terrace culture.

Lastly, the development of terrace landscapes in winter needs to be strengthened. Currently, only half of the year is spent farming, with the other half mostly idle. We should explore ways to develop winter landscape resources that allow tourists to enjoy different scenery in different seasons. Anyway, in the future, tourism development should still be guided by the government and led by enterprises, and involve cooperation from villagers.

A view of Guilin's residential changes from an architectural designer

"Rivers Flowing Silkily like Bluish-Green Belts;
Mountains Standing Tall as if Jasper Hairpins.
The Locals Collecting Exquisite Feathers of Kingfishers;
While Cultivating Aromatic Oranges for Their Livelihood."

by Qing Zhishan

Although my ancestral hometown is located in Jiangxi Province, I grew up in a small town nestled within Guilin (the Capital city of Guangxi Zhuang Autonomous Region). When I was a teenager, I got caught up in the epoch-making movement of "Youngsters Relocating to the Countryside". In the rural youth settlements, young people from all corners of the nation converged to work on farms, make fire and cook meals together, and live a bustling communal life. Back then, living conditions were simple and crude: we would wake up before dawn every morning for ploughing, weeding, and fattening animals—day after day for six consecutive years. I loved singing and often organized singing activities in fields and mountains during my spare time. Additionally, I played the *erhu* (*translators' note: Chinese*

Together with My Wife

2-string fiddle) and accordion while touring around villages to perform art shows. Fortunately, during this period of time, I had the pleasure of making acquaintance with my beloved wife, who has been a constant companion throughout my entire life.

In 1972, the local people's commune requested production teams to build granaries, and I made great efforts to accomplish the task. So, the commune sent me to Guangxi Architecture School to pursue a major in village construction. I had to learn drawing, construction, engineering and management from scratch. Upon graduation, I returned to the commune and worked as a part-time construction technician for farmhouses.

In October 1975, I was transferred to Guanyang County Construction Company in Northern Guangxi. Later on, I joined the county's architectural design office to become a grassroots technician and actively participated in the constructions of my hometown. In the early 1980s, I successfully passed the adult college entrance examination and enrolled at Chongqing Institute of Civil Engineering and Architecture for systematical studies.

Photos Taken during College Study

In 1993, I was transferred to a design institute in Guilin and became the head of the architectural design department until 2023. With over three decades of experience, I have continuously honed my skills on the front line. Whenever confronted with a new construction type, I attentively listen to the experience shared by predecessors, actively cooperate with colleagues from other majors, take initiative to visit construction sites to learn about emerging technologies, and pass on my knowledge and experience to younger generations.

Training on Traditional Residential Buildings in Northern Guangxi

As an architectural designer, I have learned the construction techniques of traditional folk houses in Northern Guangxi and continuously refined my skills for designing rural dwellings. I have personally witnessed the transformation process of Guilin's architectural styles and urban appearances and wholeheartedly dedicated my passion to the city's construction.

An Initial Glimpse of the Traditional Residential Features in Northern Guangxi

"Rivers flowing silkily like bluish green belts; Mountains standing tall as if jasper hairpins." (*Translator's note: this verse is excerpted from a poem depicting Guilin and was written by Han Yu, a historically famous poet and politician in the Tang Dynasty around 1,200 years ago.*)

This is one of my favorite poems enchanting Guilin. When discussing the distinctive traits of residential houses in Northern Guangxi, we must commence with the lands of Guangxi.

Boasting a long history and rich culture, this region was once inhabited by the Liujiang Man and Qilin-Mountain Man some forty to fifty thousand years ago. Following Emperor Qin Shihuang's unification of ancient China, he initiated construction projects that successfully linked the Pearl River system with the Yangtze River system, resulting in 70% of water flowing into the Xiangjiang River and the remaining 30% into the Lijiang River. The enduring influence of Central

Plains culture and Lingnan culture on Guilin has left an indelible mark on its folk history, evident unique architectural features such as cornices adorned with curved horns, Ma Tau walls (*translator's note: a king of traditional pediment*), as well as overhang buildings called Qilou for shops and residences. However, due to its geographical isolation and inconvenient transportation, Northern Guangxi embraced Central Plains culture much later than other regions did. Despite having a relatively small number of aboriginal residents, they have managed to preserve their traditional ways of local lifestyle. Geographically characterized by mountains in this region, stilted buildings are commonly found among folk houses along with cultural characteristics from Dong, Yao and other minority nationalities.

My Manuscript: Wind-Rain Bridge and Kitty Heights

Northern Guangxi boasts a picturesque karst landform in mountainous areas, characterized by undulating hills and limited flat terrains, adorned with lush fir and bamboo forests. In terms of architectural forms, the local residential buildings are primarily constructed along hillsides, using fir as the major materials for wooden stilted houses with dwellings on the second floor gracefully extending over the bottom space. The foundation is typically crafted from bluish-gray stones, while the main structure is made of fir. With an elegantly covered roof of small black-gray tiles, these black-gray buildings seamlessly blend into the surrounding natural landscape. The building layouts vary between three or five rooms, creating a harmonious skyline that ascends and descends in an orderly manner. These traditional buildings showcase exquisite elements such as overhanging gable roofs, draped eaves, cantilever platforms, stilted structures layered upon one another, cantilever corridors, intricately carved columns known as Chui Hua Zhu, carved window frames and the like. The patterns are elegantly smooth-lined, imbued with profound and auspicious meanings and vibrant in color, exuding an air of simplicity yet refined elegance.

The construction and renovation of traditional residential houses in Northern Guangxi still adhere to the traditional craftmanship of artisans and uphold local customs. The rituals are conducted with grandeur and fervor. When a house is being built, the whole village is mobilized. Taking the girder-placement ritual as an example, it involves several stages such as logging, log selection, girder making, offering ceremony, girder placement, girder decoration, throwing the special "girder rice cakes", and expressing gratitude towards the girder. Each stage is executed with piety, yet in a lively atmosphere, ensuring a secure and proper placement of the girder while bestowing blessings upon the abode for its enduring existence.

Refining the Skills for Designing Rural Residential Buildings

In the 1970s and 1980s, Northern Guangxi was primarily an agricultural region, with impoverished living conditions. Villages lacked of fundamental infrastructure such as roads or electricity, while people resided in dilapidated huts. There were few professionals knowledgeable in architectural planning and design at that time. Technicians like myself had to do whatever was needed: devising plans, drafting blueprints, designing buildings, working out structural construction drawings and even basic water and power systems—all of the work was learned on the job.

Gradually, subtle changes occurred with the villagers' way of life: the villages were supplied with electricity, which facilitated agricultural and sideline production; meanwhile, roads were constructed for easy product delivery. I recalled that in 1985, a "Farm House and Cultural Center Design Competition" was organized, during which esteemed predecessors and architects, Mr. Huang Rifu and Mr. Yuan Huaishan, meticulously mentored me and encouraged my participation in the competition. At that time, my relatives were engaged in the chicken hatching business, so I often went to lend a hand. Hence, I designed a "Hatching House", which won the second prize in Guangxi Zhuang Autonomous Region in 1985.

Later on, I collaborated with my classmates on a project called "Guanyang Dongjing Cultural Center", which won the third prize in the autonomous region. The two designs were included in a book titled *Designing Works Compilation of*

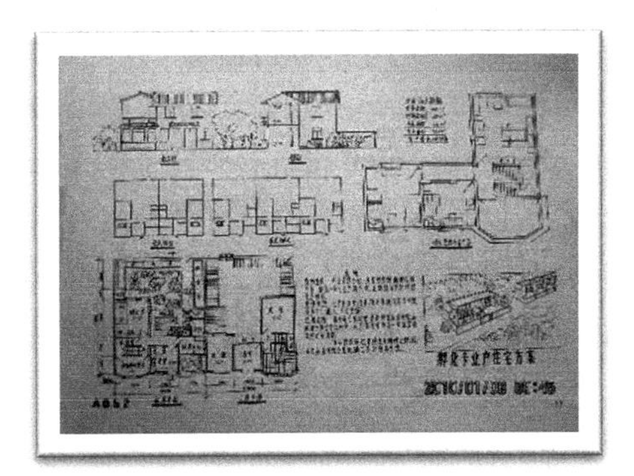

奖 状

卿志山同志：

在一九八四至一九八五年度全区农村住宅及集镇文化中心设计竞赛中，荣获农村住宅二等奖特发奖状，以资鼓励。

广西壮族自治区城乡建设委员会
广西壮族自治区土木建筑学会
一九八五年五月

The Hatching House Design Drawing and Award Certificate

Rural Housing and Township Cultural Centers of Guangxi Autonomous Region, and my name was honorably listed in the catalog alongside those of famous designers in Guangxi's design industry. How encouraging!

奖 状

卿志山、黄日孚同志：

在一九八四至一九八五年度全区农村住宅及集镇文化中心设计竞赛中，荣获洞井文化中心三等奖特发奖状，以资鼓励。

广西壮族自治区城乡建设委员会
广西壮族自治区土木建筑学会

Design Drawing of Guanyang Dongjing Cultural Center and Award Certificate

Witnessing the Transformation of Guilin's Urban Appearances

Since 2000, Guilin has entered a period of rapid urban transformation and witnessed numerous construction projects in which architectural design institutes from all over the country have participated. Simultaneously, academic lectures were held inviting renowned architectural masters, such as Wu Liangyong and Zhang Kaiji. As an expert in project evaluation, I was selected to assess several of key projects, including the Central Square, the renovation of Zhongshan Road, as well as the planning and construction of the scenic areas for two rivers and four lakes in Guilin.

During that period, there was a prevailing trend towards grand European architectural style, which led to frequent utilization of European architectural elements in new constructions and renovations. However, despite its grandeur and gracefulness—whether it be a single building or groups of them —this style appeared incongruous with the local landscape features of Guilin, which are characterized by softness, exquisiteness and agility in its surrounding mountains and waters.

After years of study and practice accumulation, together with other insightful experts, I have summarized two crucial orientations for the development of architectural design in Guilin. These orientations emphasize the traits of the local landscape and culture: one is to harmonize with Guilin's natural environment and scale; the other is to incorporate the essence, elements and symbols of traditional residential buildings in Northern Guangxi into architectural details.

My Design Work and the Location Shooting:
Complex Building for Qianjia Cave Scenic Area of the Greater Guilin Area

Aids to Younger Generation Architects

When I reached retirement age, I accepted the reappointment invitation from Blue Sky Technology Co., Ltd. and became the vice president in charge of architectural design business. With this opportunity, I would like to devote my remaining life

My Design Work：the Water Street of Yao Ethnics of Huajiang

to the continuous endeavors and accompany younger generations on their career paths.

The company annually organizes learning activities focused on new technologies, where I actively engage with young designers in acquiring emerging technologies. I am delighted to share my experience with younger generations wholeheartedly and without reservation.

Since the initiation of rural revitalization by the authorities in Guangxi Zhuang Autonomous Region, my involvement in rural renovation projects has significantly increased. Leading a team of young architectural designers, we are dedicated to achieving a pristine and aesthetically pleasing appearance of the countryside while simultaneously enhancing the functionality of rural facilities. Our goal is to establish a "harmonious rural community that integrates ecological balance, improves living conditions, and enhances productivity".

Based on my years of work practice, the typical architectural layouts of self-built rural houses in southern China in recent years can be summarized by the characteristics of "extensive ground-floor space", "stilted middle-level space" and

'extensive ground-floor space', 'stilted middle-level space' and 'reduced upper-level space' presented in rural houses of Southern China

"reduced upper-level space". In order to address the issue of thousands of villages having a monotonous appearance, I aim to gain recognition by sharing common sense and knowledge with villagers. Consequently, I have developed several sets of design schemes for renovating facade features in rural communities. Moreover, I

Design Drawings of an EPC Project by Blue Sky Company: Ancient Street Renovation of East-Guangdong County Gild

Design Drawings of the Rural Residential Renovation Project of Chang'er Village, Jinxiu Yao Ethnic County, Guangxi Zhuang Autonomous Region

personally conduct demonstration and training sessions to tackle this problem. I am determined to share my experience with young designers on how to renovate rural residential houses in Northern Guangxi.

Looking back on the past, whether driven by personal passion or the needs of our country, I have embarked on a fulfilling journey from being a young student to

Self-portrait

becoming a builder, then an architectural designer, and finally a mentor helping train young architects. I sincerely hope that younger generations will achieve even greater accomplishments in the future, just as the ancient Chinese proverb goes: "Originating from blue but surpassing it with bluish green", meaning that worthy disciples excel their masters.

My Hometown

by Tang Xinyi and Qin Jiangxia

My hometown is in Dejiang County, Tongren City, Guizhou Province, where I actually have two homes. One is located in a small village nestled among lush mountains and clear waters where nature's gift abound while the other is situated in a more modern small county still surrounded by mountains but seemingly disconnected from the outside world.

When my children were young, I used to take them and my mother to travel by bus to my hometown every weekend and during school holidays. The buses were

Queuing at the Bus Station

A Village Surrounded by Mountains

always packed with belongings and people who had to compete for a place to stand while chatting loudly with acquaintances they happened to meet.

In the early morning, as the first rays of sunlight pierced through the dense foliage and fell upon the land of the village, a golden glow bathed the entire village. The villagers rose early to commence their daily tasks. The young and strong rolled up their sleeves and wielded their hoes, tilling the fertile soil. Their sweat glistened in the morning light, and their figures appeared to merge with the earth underneath. Meanwhile, hardworking women gracefully glided through wheat fields, wielding their scythes to harvest golden grains. Their movements were deft, each stroke of the scythe being a prayer for a bountiful harvest. Even the young children did not lag behind; some followed their parents into fields to assist in gathering crops into baskets while others played nearby with small shovels digging into dirt. Though their contributions may have been small, they added a touch of joy and innocence to the fields. On this land that was not particularly fertile, men, women, and children worked together harmoniously like a symphony of life itself—laboring not only for sustenance but also out of love for this land and passion for life.

In the early 2000s, the mountainous villages in Guizhou exuded a simple and natural atmosphere, contrasting sharply with the noise and bustle of modern urban life. During this time, houses in the villages were primarily constructed using rammed earth, bamboo and wood materials. They also featured simple furniture and household items. Only wealthy families could afford brick and tile houses, which were admired by all villagers for their black-and-white televisions.

My home was a wooden house. Compared to rammed earth, wooden houses provide stronger structural integrity and greater durability by using local wood from the nearby mountains. The roofs were often covered with wooden tiles or stone slabs to prevent rainwater from seeping in. Inside, there were handmade wooden tables, chests, bamboo beds and other items crafted by family members. During this time, the types and quantities of furniture were limited; most families made furniture by themselves based on their actual needs. In addition, we also possessed some basic household items such as earthenware utensils, bamboo baskets, and wooden kitchen utensils —simple yet durable, all made from local materials.

Despite the harsh conditions, villagers were able to create warm and beautiful homes through their wisdom and hard work. During every festival, each household would hang red lanterns and set a table of delicious food outside the door, inviting relatives and friends to gather together and spend a joyful time.

Making Pumpkin Fritters on New Year's Eve of 2024

I often recall the scenes from my childhood. It was challenging to attend school in the countryside, but my parents arranged for my two brothers to take care of me so that I could receive education earlier. On the way to school, my brothers took turns carrying me on their backs. Due to my mischievous nature, I would often insist on walking alone. However, my eldest brother would always silently accompanied me, while my second brother sternly insisted on carrying me all the way home. Some other family members mistakenly thought that my brothers were bullying me, but they didn't realize that this was their unique way of expressing love and care.

The First Family Portrait

As time passed, my brothers grew up and transitioned from a mountain primary school to a town junior high, while I grew from being a little kid into a teenager who could take care of myself. I started learning how to look after my younger siblings. Despite our family not being wealthy, our parents' hard work ensured that we had enough food and were warmly dressed, especially thanks to my mother's selflessness and dedication which supported the entire family.

Later on, young people gradually left the village to work elsewhere, and some even ventured further afield. When they returned home with their pockets full after a year of hard work, more young individuals in the village looked up to them with admiration. Consequently, during that time, many junior high school graduates followed in the footsteps of their elders and sought employment outside the village. I was no exception and embarked on a life-changing journey following my neighbor's elder sister. Through hard work and government support, my husband and I were able to purchase a house in the county town. As a result, I have had my second home ever since.

Family life in the countryside is notably different from that of the county. Firstly, rural homes lack access to tap water, which poses challenges for daily life. However, in urbar area, infrastructure such as tap water and telecommunication has been widely adopted earlier. Secondly, while people in villages relax by randomly choosing a household courtyard to sit on chairs and chat leisurely, those living in the county have more options for passing their time, such as watching TV at home or enjoying the latest movies when a touring film team visits the community. The most significant difference lies in education; few schools are located in rural areas and they are often too far away from home. On the other hand, schools in the county are much closer and more convenient.

Over time, my hometown has undergone a dramatic change.

The development of society has brought about changes in lifestyle. With the promotion of urbanization and the convenience of transportation, people's lifestyles have significantly transformed transitioning from the agricultural era to a more diversified and contemporary way of life. In the past, rural residents primarily relied on farming for their livelihoods, leading relatively simple lives. However, with the acceleration of urbanization and advancement of information technology, rural residents' lives are becoming increasingly diverse and focused on personalization and quality.

I am very grateful to our country and government. I recall a day in 1997 when electricity was brought to the mountainous countryside. The entire village was filled with excitement on that day when electricity was turned on. At that time, nobody knew that using electricity required payment, thus lights were left on all night long. Even some elderly people remarked that witnessing light without oil in their lifetime was priceless, and they wished for it to be kept on all the time.

The policy changes have also had a profound impact on people's living and working environments. The government has implemented a series of policies to support rural development, including agricultural subsidies, infrastructure construction, and cultural and tourism promotion. These policies not only promote

The Family Portrait of Year 2021

the economic development in rural and county areas, but also improve people's living conditions and working environments. With the development of rural economy and government support policies, working conditions have improved, income has increased, and peasants have more employment opportunities. They are no longer limited to hard labor in the fields but can achieve their dreams through various means. As a result, many young individuals now choose to stay in their home villages and start businesses, achieving their career dreams by developing agriculture and tourism.

Furthermore, the advancement of technology and increased awareness of environmental protection have also brought about significant changes to our lives. In the past, our contact with the outside world mainly relied on letters and telephones. Now, with the help of the internet, we can easily communicate with people all over the world and obtain various information and knowledge, which undoubtedly greatly improves our quality of life. In the past, due to a lack of awareness regarding environmental protection, there were problems such as environmental pollution and resource waste. However, with societal progress and increasing awareness of environmental protection, remarkable results have been achieved in terms of environmental preservation and management. The quality of air, water, and soil has improved significantly as well, making people's living environment cleaner and more comfortable.

The Water Dragon Festival of Dejiang County

The most beneficial thing for my family is education. The government has significantly increased its support for education and made improvements to educational conditions. New schools have sprung up in the county town like mushrooms, each one surpassing the other in quality. Moreover, rural areas have also received enhanced government support to construct more schools capable of accommodating a greater number of children's aspirations to soar towards a brighter future and lead a happy life.

Additionally, social and cultural changes have profoundly impacted my life. The progress and opening up of society, rapid economic development, accelerated urbanization, increased population mobility, and narrowing down urban-rural disparities, the connection between urban and rural areas has become closer. As a result, we have more opportunities to interact and communicate with each other, allowing us to learn more and broaden our horizons. This cultural exchange not only enriches our lives but also fosters confidence and open-mindedness within us. Moreover, the changes in social structure have brought forth both opportunities and challenges that I am increasingly concerned about regarding the direction and trend of social development.

Finally, changes in values have also had a profound influence on my life. With the progress of society over time, people's values and beliefs are also evolving as well

The No. 1 Middle School of Dejiang County, Guizhou Province

as gradually replacing traditional values with new ones. Nowadays, people place greater emphasis on individual development and humanistic care. This change has not only affected my attitude towards life but also enabled me to gain a deeper understanding of the diversity and complexity of society.

The implementation of various policies, economic development, and social changes have resulted in my hometown, small county town, presenting a characteristic integration of modernization and traditional culture. Compared to the past, the home environment has become increasingly diverse and comfortable with significant changes in terms of decoration, facilities, and lifestyle. My home boasts spacious and bright living rooms, bedrooms, and a kitchen; it is even equipped with modern household appliances such as a TV, refrigerator, and washing machine. The types and quality of furniture and household items have also improved. Traditional wooden furniture is gradually being replaced by panel furniture or solid wood furniture while various decorative paintings and photo walls appear on the walls. What delights me even more is that there are large supermarkets near our home which provide great convenience for trade and shopping. In the evening, my family can take walks along the riverbank park where the intimate atmosphere gradually warms up as we leisurely stop along the way; for young people, the booming commercial streets, cinemas, and gymnasiums provides them with a variety of choices for leisure time activities as well as opportunities to make friend.

These changes not only make our lives more convenient and comfortable, but also transcend material improvements, providing spiritual satisfaction and self-confidence, enhancing our quality of life and sense of happiness. We used to live in poverty; however, through education and the support of government policies, we now have more opportunities and choices to enrich our lives. It is easy to imagine that for children who were raised in environments lacking infrastructure, these changes represent endless possibilities and hope for them.

Compared to the harsh conditions we used to endure, our living conditions have improved significantly. However, there are still some areas that can be further enhanced.

Firstly, I believe that the most lacking aspect is infrastructure improvement and expansion. Despite significant government investment in improving infrastructure, such as roads, tap water, and electricity in recent years, there are still areas with inadequate infrastructure, especially in remote or poverty-stricken region. This situation adversely affects residents' quality of life.

Secondly, safety hazards and environmental pollution issues exist in the living environment. For instance, some areas have substandard construction quality and potential safety hazards that threaten residents' lives and property. Moreover, effective management and treatment of industrial pollution and exhaust emissions are necessary to protect residents' health and ensure sustainable eco-friendly environmental development.

Community construction and management also require improvement as part of the living environment. A well-maintained community environment promotes communication and interaction among residents, enhancing social cohesion and a sense of belonging. Therefore, it is necessary to strengthen community facilities construction and improve community management to create a safe, harmonious and livable environment.

Moreover, convenient transportation is crucial for a well-qualified living environment. A well-developed transportation network facilitates residents' travel, promoting economic development and social interaction. Therefore, it is necessary to enhance urban traffic planning and improve the coverage and service level of public transportation systems to reduce traffic congestion and environmental pollution.

In summary, despite significant improvements in the living environment, there are still issues such as incomplete infrastructure, safety hazards, environmental pollution, and inadequate public services that require further efforts from the government and all sectors of society to increase investment and to enhance regulation and improve management. These efforts will ultimately elevate residents' quality of life and sense of happiness for residents.

Finally, I would like to address the fact that changes in social life reflect China's social development and are the result of our collective efforts. These changes have given us a deep appreciation for progress in social development, particularly in education, healthcare, and infrastructure. These unforgettable transformation not only make me cherish my present life more but also inspire me to work harder towards future development. In my heart, Dejiang County of Tongren City is not just a place name but a beautiful home that witnesses the diligence, simplicity, and endeavors of the Chinese people. As a young Chinese born during the era of reform and opening up, I will dedicate my life to my hometown and my motherland like my predecessors did by contributing to realizing China's dream for national rejuvenation.

The Transformation of My Campus Life

by Wang Changliu

The Primary School Campus in the 1980s

I was born in Hainan Province, the beautiful and grand island renowned for its abundant natural resources. Ever since I could remember, my brother and I had accompanied our parents from the countryside of our birth and moved to Haikou City (*translator's note: the capital city of Hainan Province*), where we resided in the compound of the Farm Reclamation Supply and Marketing Company (hereinafter referred to as "the Company"). This compound boasted not only office buildings, but also staff dormitories, cafeterias, warehouses, a yard for vehicle teams, a post office, shops, basketball and volleyball courts along with all other essential facilities. The verdant surroundings were enchanting with coconut trees adorning both sides of the roads while a dedicated park within the compound flourished with lush fruit trees, such as star fruit, mango, jackfruit, lychee, longan, wampee and the like. In my memory, aside from attending school and doing homework, "planning" how to secretly pluck fruits without being detected by the logistics personnel was the most important thing for me and my friends.

The primary school was the utmost crucial public service in the compound for children. Although it was a municipal primary school, most of the enrolled pupils were children of the Company employees. Therefore, it could be considered as a dedicated primary school for the Company. The primary school covered an area of about one hectare. Compared to now, schools in the late 1980s were relatively simple. The campus consisted of five buildings: the main building opposite the school gate, teaching buildings, staff dormitory and a logistics building spreading along both sides of the main building, providing functions such as campus shops and sporting goods warehouse. The main building and staff dormitory had multiple floors while others were merely one-storey bungalows. These five buildings formed a "U" shape enclosing a playground for students activities. The playground was actually an unhardened field without running tracks, which resulted in my first shoulder fracture when I collided with a classmate during PE class. The facilities and conditions on the primary school campus merely met the basic teaching requirements at that time. Since there were no landscape facilities on the campus, the park and orchard in the company compound became a tutoring base for the students' practice courses.

It took me less than five minutes to walk from my home to the classroom. I could even say that I had lived in the primary school for six years and had a wonderful childhood.

The Middle School Campus in the 1990s

I attended both middle school and high school at the same institution which covered an area of approximately 10 hectares. It accommodated around 2,500 students from two departments within a unified compound: middle school and high school. It comprised two main teaching buildings, each three floors high, that housed classrooms for each grade level on the same floor. The southern building

Scenery of Hannan Overseas Chinese Middle School

Dormitory Building of Hainan Overseas Chinese Middle School

Track and Field Stadium of Hainan Overseas Chinese Middle School

catered to the three grades of the middle school department, while the northern one served those in the high school department. The campus boasted a picturesque environment adorned with boulevards and delicate gardens. In comparison to my primary school campus, this institution offered an array of additional facilities such as libraries, a standard track and field stadium, students' dormitories, cafeterias, faculty communities, etc., effectively catering to the basic needs of nearly 3,000 teachers and students in terms of studying, living and working. During my six-year stay at this school, consistent improvements had been made to its infrastructure starting from public spaces onwards. I witnessed renovations being undertaken on the track and field stadium: changing its orientation from original east-west to north-south according to standards; adding a podium; leveling the running tracks; although they still consisted of a mixture of cinder and loess soil. Additionally, concrete basketball courts were successfully constructed as well. However, regrettably, my favorite football patch where I played most often remained without grass. The year I graduated from highschool also witnessed renovation taking place in our student cafeteria.

As a day student, it took me no more than 15 minutes to walk from home to school. During noon time, I could go home to take a nap. After supper, I would cycle back to school for evening self-study. Except for the summer and winter holidays, I usually spent more than 10 hours at school every day . The campus truly held the memories of my middle school time.

The College Campus at the Turn of the Century

In the early decade of this century, I completed my undergraduate, master's and doctoral degrees at the same university. The facilities and conditions on the university campus were much better than those in the middle school, not to mention that it was the most esteemed agricultural university in China. What impressed me most was that our campus boasted exclusive spaces such as experimental farming fields, pet hospitals and a rugby field. From middle school to university, the primary challenge for me was adapting to dormitory life. During my undergraduate years, I lived in Dormitory Building Ⅴ which hosted 6 students in one bedroom with bunk beds. Students pursuing a Master's degree lived in Building Ⅶ with 4 students sharing one bedroom which was also furnished with bunk beds; while Doctoral students lived in Building Ⅲ , where 2 students shared a dormitory suite with an independent kitchen and bathroom. The per capita area increased from 1.5 square meters for undergraduates to 6 square meters for doctoral students. In terms of studying environment, every university had been developing vibrantly over the past ten years, and the hardware facilities for teaching and research activities had been constantly improved. During graduate studies stage, my tutor arranged independent working desks for us. In addition, we were allowed to use our team studio located in Shennei Mansion, which meant that we were no longer confined solely to occupying seats for study and research within public classrooms or libraries. In my recollection, every time during the final review stages of each semester, it would be difficult to find available seats at No. 174 Classroom of the Earth Chemistry Department; lamps would be lit throughout nights. Anyway, the campus facilities and environment had significantly enhanced over the decade including a newly built student cafeteria named "Yiyuan", student dormitory buildings, tennis courts, public bathhouses, Resources and Environment Building, Life Science

The Old Teaching Building in the West Campus of China Agricultural University

The Dormitory Building in the West Campus of China Agricultural University

Building, Animal Hospital, etc. The environment around the campus had also been continuously improved with the addition of hiking trails to Baiwangshan Forest Park, the renovation of the Jingmi Canal, the transformation of Malianwa Market into a commercial complex, clearance of a large number of illegal building clusters in Xiaojiahe community, and opening of Anheqiao North Station on Metro Line 4... Outside the Fifth Ring Road in Beijing, the transformation of No. 2 West Yuanmingyuan Road is a vivid microcosm of the rapid development that our capital city had experienced in the early decade of this century.

The University Campus in Southwest China

After obtaining doctoral degree, I ended my life as a "Beijing drifter" and moved to Chengdu City. I once worked in the planning management department of the local government for two years. After that, I transferred to Southwest Minzu University, where I have been working as a teacher for over 10 years. I have resided in the faculty community during the past decade. The essential reason for my choice is the convenient educational conditions provided by the university: an affiliated kindergarten available on the campus and a primary school within a 10-minute walk. Children could easily access campus squares, small gardens, sports fields and the like after school. The safe, friendly and beautiful environment of the university campus is very beneficial to the growth of children. Especially during

Landscape of the Wuhou Campus of Southwest Minzu University

The Dormitory Building on the Wuhou Campus of Southwest Minzu University

The ethnology Museum on the Wuhou Campus of Southwest Minzu University

the past three years of the Covid-19 pandemic, the benefits has been particularly obvious. The university has properly implemented epidemic prevention and control measures, enabling children to play on campus instead of being confined at home. Backed by the campus cafeteria, we never snapped up any food items in those days. Now, occasionally I take my child to do homework at the school library. Actually, the Wuhou Campus of the university is situated in the central urban district of Chengdu City and requires to maintain a stable planning layout primarily focused on improving functions and quality. This includes redesigning and decorating the

Landscape of the Airport Campus of Southwest Minzu University

Pomegranate Plaza on the Airport Campus of Southwest Minzu University

Ethnology Museum, installing elevators for residential buildings in the faculty community, and repairing two municipal cultural protection buildings.

Since 2015, my workplace has gradually been relocated to the Airport Campus of the University. Despite being put into use as early as 2003, the construction speed has significantly accelerated in the recent decade. The campus now boasts newly constructed facilities such as Jingwenyuan Experimental Building, Innovation and Training Center, Engineering Training Building, Food & Memory cafeteria, Experts Center, International Students Building, an indoor gymnasium, outdoor basketball courts, and a kindergarten among others. One of its most notable achievements is the completion of Pomegranate Square, a landscape square constructed right above the cover section of a tunnel that connects the northern and southern campuses into one unified campus which used to be separated by municipal roads. This project greatly facilitates the study, work and daily life for teachers and students alike. As a full-time teacher specializing in landscaping architecture, I had the honor to represent Party A and contribute to part of the landscape design for this central square. In this decade, many colleges and universities have successively developed new campuses outside the central urban areas, and these new campuses have become focal points for construction.

Over the course of four decades, spanning from adolescence to middle age, it appears as though I have never truly departed from the campus environment. I have borne witnessed the transformations across various campuses in several cities. Presently, there exists no inclination within me to relocate away from this cherished environment, for I eagerly anticipate the dawn of a new digital era that will bring about a fresh appearance for the campus environment.

Human Settlement Impression

by Wang Guangzheng

In 1963, I was born in the Wang's Courtyard of Shicongtou Village, Jincheng City, Shanxi Province, China. Surrounded by four mountains—Longtou (Dragon-Head) Mountain, Fengtou (Phoenix-Head) Mountain, Zhutou (Pig-Head) Mountain, and Yutou (Fish-Head) Mountain—Shicongtou village iswell concealed in a covert valley. The village enjoys sound security with Changhe River flowing around it as a natural moat that prevents bandits and robbers; it is easy to defend but difficult to attack. One can effortlessly to see the mountains but not the village, or hear voices without seeing people. Compared with other ancient castles in the Qinhe River basin, Shicongtou Village has neither high walls nor watchtowers; instead, it is a purely natural ancient village that seamlessly integrates with and is well protected by the surrounding natural environment.

Shicongtou Village has preserved ancient stone carvings known as "Stone Path through the Clouds" and magnificent earthen-wall fences named Village Kulve, along with over ten ancient public architectures. Religious sites include the Grand

Part of the Village

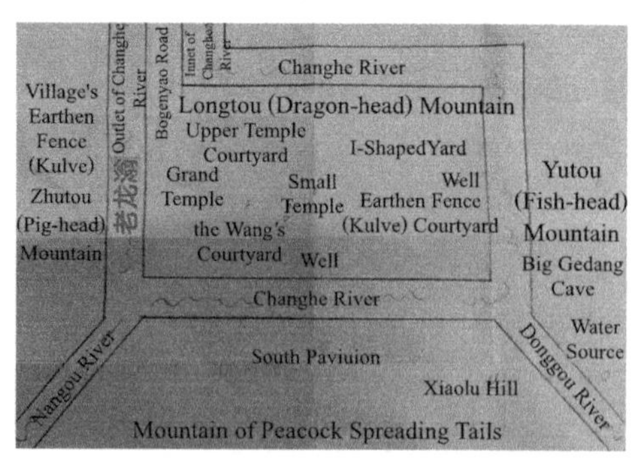

Terrain and Landscape Map of the Village

Village Kulve (Earthen Defence)

Kulve Yard

Temple, the Small Temple, and the Southern Pavilion. Ancient architecture facilities were built in Ming and Qing dynasties including Upper Temple Courtyard, West Courtyard, Shangtou Courtyard, the Wang's Courtyard, I-Shaped Yard, Back Yard, Chessboard Yard, Screen-Wall Yard, and Kulve Yard.

The Village was supplied with free tap water some 50 years ago. Currently, every household has access to electricity and gas supply, and the dirt roads have been transformed into paved streets. In 2014, the Village was included in the third batch of Chinese Traditional Villages. Presently, there are two shuttle buses available daily for tourism.

The Wang's Courtyard is a building of Qing Dynasty with a history of over 200 years. It is a typical quadrangle dwelling in northern China, featuring four main halls with eight small chambers.The upper floors have extended eaves and wooden

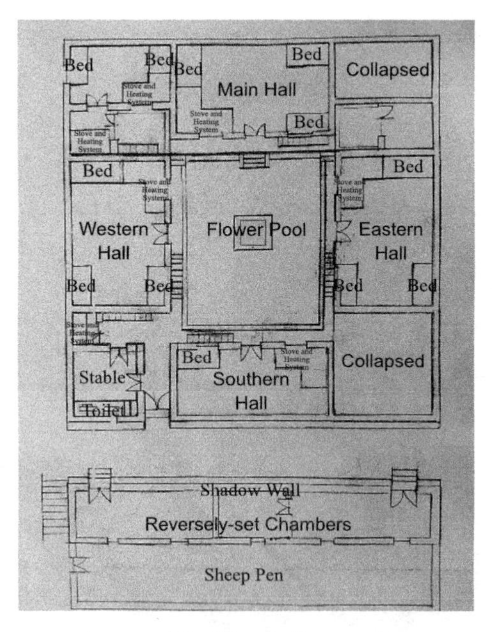

Layout of the Wang's Courtyard

stairway entrances, and the greenish stone steps are installed in the yards. Toilet facilities are located at the southwest cornersin the courtyards behind the main gates allowing for toilet use within the courtyards while positioning the excrement disposals outside.The main gates open to the south with rainwater drainage underneath.

My family resided in the Northern Hall, the main hall. In front of it, there were three sets of greenish stone steps (three levels) installed at the eastern, western and middle positions respectively. The middle set was equipped with revetments on both sides. There were stones used for building foundations, mounds, steps, and slope protection. Furthermore, there were stone stairways, stone window sills, stone thresholds, stone beams for the toilet, stone blocks for the feces outlet outside the gate, and greenish stone bars for paving the path outside the gate. All these stones were artificially carved to be flat and smooth with clear-cut edges. Over time, they became even smoother and brighter due to continuous grinding.

The construction materials included stone bricks for walls, wooden beams, and roof tiles. The main hall consisted of three rooms, two windows, and one door. The wooden square windows were sealed with pasted paper and decorated with

West of the Wang's Courtyard

East of
the Wang's Courtyard

Greenish Stone-bar
Path

traditional handmade art paper-cuttings by villagers which are renewed annually during New Year celebration. Stone door-mounds and sturdy wooden doors could be found there. With door-bolt inside and a latch plate outside, the door with an ornately designed head could be locked from either side. Inside the rooms were corner stoves, beds, boxes, large cabinets, small cabinets, long tea tables, square dining tables, chairs, and other daily supplies. The rooms generally served as both living rooms and bedrooms. Typically upstairs was an altar table placed with a memorial tablet bearing inscriptions of "Heaven, Earth, Emperors, Ancestors and Masters", dedicated to worshiping the immortals and saint masters. The remaining space primarily served as a warehouse for food reserves along with miscellaneous household items.

The northwest corner of my home had two additional rooms, one upstairs and one downstairs. The room at downstairs was equipped with a water tank, a stove,

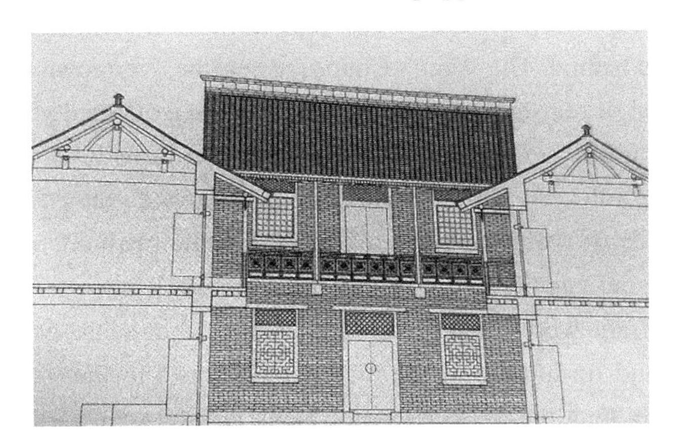

Main Hall Front of the Wang's Courtyard

kitchen cabinets, kitchen utensils, and beds. The kitchen and resting space were combined. The room upstairs served as a warehouse. Both upstairs and downstairs rooms had wooden doors and windows installed on the same wall to prevent drafts. The door was placed on door-mounds to avoid wind sweeping through it. The deep yard design created a wonderful wind shelter that naturally adjusted the environmental temperature, making the rooms warmer in winter while cooler in summer. The Wang's Courtyard accommodated four families with similar structures, layouts and functions in their dwellings. With over 20 residents living in close quarters, they frequently visited each other's dwellings for casual chats and mutual assistance. How lively!

Coal and carbon were used for heating and cooking, water was drawn from the spring in the village, and lamps were lit with kerosene or calcium carbide until electric lamps became available later. My family had a radio, a TV, a bicycle, and a sewing machine. There were only two sets of sewing machines in the village at that time, and ours was one of them. It was a tool that belonged solely to my eldest sister since other villagers did not know how to operate it. The last month of the lunar calendar was the busiest season for my eldest sister because she would sew all the New Year's clothes for the villagers free of charge, regardless of whether they were adults or children. Besides, she had to cover the cost of threads. This task required time, keeping my eldest sister busy from early morning till late night, so she was very tired during that period.

Around the year 1990, people who lived in the courtyard built their new houses and gradually moved out. As the children grew up, they also left one by one, leaving an empty courtyard behind. The Wang's Courtyard became overgrown with weeds and desolate. In 2021, a rainstorm that hadn't been seen in a hundred years struck, tore apart the building and yard and only ruins standing there now left.

Around the year 1985, my family built seven bungalows to the east of the Wang's Courtyard. Three of them were located in the middle position, while two were on the left side and another two were on the right side. They also had stone-brick walls, wooden frameworks, and tiled roofs, along with rammed-earth gables. The indoor layout and furniture placement remained almost unchanged from the old courtyard's style, with the only difference being an exclusive yard for my family alone. There was no need to share the yard with other families, which allowed for

Rural Bungalows

more convenient access and privacy.

When I was in high school at Zhoucun Middle School, the student dormitory consisted of a row of bungalows, with each room built around a circular brick-bed platform. At the edge of the platform, students' wooden box were placed to store personal items and block the wind from entering through the door. The dormitory was simple and crude. The toilets and washrooms were all located outside, far away from the dormitory, which was very inconvenient.

In 1981, I started my career and became a worker at a resin factory. The bachelor dormitory provided by the factory was a bungalow with each room shared by three staff members. There was a stove available for cooking, but we had to prepare our own pots or pans. Since I did not know how to cook, my first attempt at making cornflour mash ended up with a pot of squishy soup full of cornflour bumps. When I got married in 1987, the factory assigned me a cave dwelling with two rooms (translator's note: cave dwellings are typical residences in the Loess Plateau region of China where holes or caves are dug into earth mountains for people to live in). The outer room served as a living room, while the inner room functioned as a bedroom. It was furnished with combination cabinets, a double bed, and a sofa and included amenities like a two-cylinder washing machine, a color TV set, a sewing machine and even a motorcycle. The kitchen was a simple shed located adjacent to the entrance with tap water, a sewer system, and an electric water heater(which was converted from an old tin bucket). We also had our iron furnace welded along with radiators installed. Apart from inconvenient toilet facilities, everything else was nice and satisfactory. At that time, residing in the factory dormitory came with many benefits such as free housing, free usage of water, electricity and coal, convenient access to hospitals, and free schooling. All essential aspects related to work, housing and daily life went smoothly. Everything was satisfied, and we led an enjoyable life that others often envied.

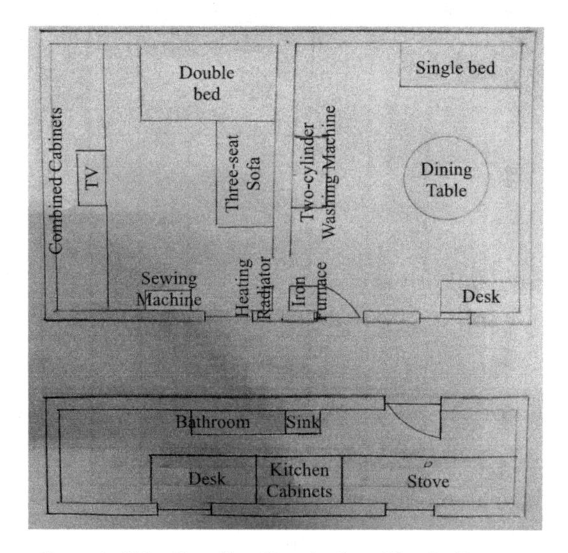

Layout of the Cave Dwelling Assigned by the Factory

With the implementation of national reform and opening up, dramatic changes unexpectedly arose, like a spring thunder shaking the earth on which we used to rely. The epochal trend urged people to liberate themselves from their old mindset rooted in the original planned economy. Everything changed: the assigned jobs were gone, the free housing system disappeared, and even free schooling vanished. How could all these problems be fixed? I used to lead a comfortable life but then felt dizzy and unable to recognize any direction in life. I lacked the ability to live independently or possess working skills. On whom could I rely? Who could lend me a hand? I could only feel hopeless, almost desperate.

The wife suggested, "Let's move to the city and find jobs there so that we can buy a place to live." With some borrowed money from friends and family, along with loans from the bank, I purchased a taxi. I was so desperate to earn money that I kept working 16 hours a day with only one meal. However, this routine continued less than a year and ended up not making money at all; instead, it cost my health, which almost killed me. I experienced serious weight gain and high blood pressure; I lost my appetite when hungry and couldn't fall asleep when exhausted. I guess I was actually experiencing Vegetative System Dysfunction at that time. It seemed like old problems remained unsolved while new ones were added in. During this period of timc, the four of my family had to rely on the support of my eldest younger brother. We stayed in his self-designed house where we occupied an upstairs room facing east. Due to space restrictions, the door and window directly faced each other. Despite sufficient natural lighting in this room, we suffered from drafts, which made it particularly cold in winter and hot in summer. In fact, my wife's feet would swell due to extreme cold during every winter season. Besides, the lack of toilet facilities upstairs made our daily routines very inconvenient.

I lived in my brother's house for six years, which was expected to be a golden period of life in terms of age and career; however, it turned out to be the lowest point of my life. During these six years, my mindset and life underwent a complete transformation as I also experienced the loss of my parents. As a Chinese saying goes, "With parents alive, life is blessed with a nestle to rest our bodies and souls on the Life Tree; but with their passing away, life becomes a lonely journey towards its end." Feeling helpless and having nothing to be proud of in life, I had an epiphany moment when I decided to prioritize living a healthy lifestyle before pursuing wealth. However, during those six years, the housing prices had doubled from a few hundred RMB per square meter to more than one thousand.

As Chinese, we believe that living in peace and harmony promises satisfactory career development. Therefore, in 2005, I sold my taxi, borrowed money (from friends and relatives), took loans again from banks, and managed to afford a new apartment for the family.

I didn't have high expectations for the new apartment; as long as it was affordable, I would be satisfied with it. During the decoration and furnishing stage, my requirements were focused on cost-effectiveness—simplicity, economy, practicality yet elegance. The apartment was not very spacious; it consisted of two bedrooms and one living room totaling 98.7 square meters on the sixth floor and had an installed floor-heating system. The purchase package of the apartment also included a pleasant basement measuring 17 square meters, which was equipped with a window. At that time, I even thought I might live in the basement during my old age. However, due to the lack of toilet facilities or access to tap water in the basement, we could only use it for storage purposes instead of living in it.

My Apartment in 2005

A few years later, my daughter got married and my son grew up while my wife and I aged. Once our son gets married in the coming years, my wife and I will need to plan for a new place to live. With this idea in mind, I had expected to purchase a small apartment in a building with an elevator to reside with my wife. In 2018, this dream came true.

My Apartment in 2018

From the large courtyard housing multiple families to the one exclusive to my family, from a room shared by two generations to an apartment for one generation, from bungalows to high-rise buildings with elevators, these are all what I have experienced in my life. I would like to write a poem to document the achievements in residential environment transformation, which I am currently enjoying, as follows:

Elevators go up and down in high-rise mansions.
Paved roads connect all the ways around.
One-yuan RMB is charged for each bus ride,
Making it affordable and convenient.
A civilized city with courtesy,
This is my beautiful hometown.
Nutritious rice and noodles, delicious fish and meat,
All are served on dining tables.
Stable earnings secure our food and clothing.
Progressing with the epochal trends of social development,
What a thriving prospect!
Thanksgiving dedicated to my dear motherland,
Happy times bring prosperity to this country and peace to our people.

Witnessing the Advancement in Living Environment of the Era

by Wu Xiangyang

Country Time

An inconspicuous small village nestled in the southern region of Henan Province, quietly lies within the long river of history, bearing witness to the growth and transformations of countless generations. In the golden autumn of 1971, I was born into this world on this plain and barren land, embarking on a childhood accompanied by soil and nature.

At that time, our village had a large population but limited land resources. Locals often referred to it as "70% of mountains, 10% of flat fields and 20% of farmlands". Moreover, due to low productivity, the land's production capacity was seriously insufficient. My home dwelling was an old rammed-earth house constructed with thick walls and wooden windows without glass but barely covered with fertilizer bags to protect from wind and rain. The roof was covered with black tiles while inside the house you could see dark purlins and rafters marked by years of smoking. This type of housing was common during that era—cold in winter and hot in summer; marks of time were engraved onto their appearance after enduring years of wind and rain.

When night fell, it was particularly quiet as there were no electric lights; only the faint but warm glow of a kerosene lamp illuminated the desk where I did my homework while my mother sewed soles for our shoes. This scene still flickers in my dim memory like a flame. Although the light of the kerosene lamp was very weak and I had to frequently adjust its wicks, I did not develop myopia at all. This light warmed my heart and ignited my desire for knowledge and curiosity about the outside world.

The kitchen, being the warmest place in the house, was equipped with a huge wood-fired earth stove that occupied the central position and had two large iron pots. Every morning, when the cocks crowed for the third time (about 5 am), my mother started her bustling day. She skillfully added wood to fuel the fire, and soon the smoke from burning firewood filled up the kitchen. The stove had long been blackened by smoke, while the bottoms of the big iron pots needed scraping every three or five days to maintain heat conductivity. The smoke curled up and weaved together with the morning mist—this picturesque scene could be captured by modern painters or photographers as typical countryside scenery. In my eyes,homemade food meant brown rice steamed in a big iron pot, mostly porridge but occasionally coarse steamed buns too. Looking back now, it was indescribably tasty and could easily fill my empty stomach—a delicacy that no appliance in a modern kitchen could replicate. However, at that time it was actually hard to swallow unless I was exceptionally hungry. Nevertheless, it was challenging for my parents to raise me with their diligent hands and simple food. Those were days worth savoring and memorable.

The Rammed-Earth House Constructed in the 1950s

Although the living conditions were simple and rough, we enjoyed the intimate contact with nature that is now difficult to find in the modern life lived in high-rise buildings. In summer, the night sky seemed to be washed with exceptionally bright stars. We often lay on a mat spread out in the courtyard, counting the stars above and listening to insects singing while accompanied by the annoying buzz produced by mosquitoes that could never be driven away. During the daytime, the sun blazed on us as we worked in the farmland holding sickles in our hands, sweating like rainwater washing through; we cut firewood up in the mountains, and each cut engraved with our passion for life. Walking through a lush vegetable patch, I handpicked a cucumber and gently wiped it with my sleeves before enjoying its crispy and refreshing flavour. That wasnature's most rustic taste, instantly bringing moisture into my heart. In summer afternoons, we often swam and frolicked in wild ponds, enjoying their tender water and freedom of joy. In winter, the chilly winds blew and our house became terribly cold due to the lack of proper heat preservation measures; therefore, we could only wear thick cotton clothes and trousers while sitting in front of a hot furnace but feeling cold at our backs, roasting sweet potatoes or baking glutinous rice cakes. Family members gathered around the furnace knitting, mending and talking. What warmth and joy we experienced! These activities made our winter days less difficult.

Over the years, with the implementation of the national "Reform and Opening-Up" policy, my hometown had quietly transformed. In the mid-1990s, I started my career life, and my brothers' income also increased. Consequently, we were able to improve our living conditions by demolishing the old house and constructing a two-story building in its place. This became the first two-story building in the village. As a result, our living conditions improved significantly with more rooms and space; furthermore, the quality of the building was greatly enhanced. We no longer worried about wind or rain in the new house. Electric lights replaced kerosene lamps, self-pressurized wells substituted conventional wells, and later on tap water supply was introduced. However, we retained the wood-fired stoves until now. With urban development taking place rapidly, fewer people chose to stay in their hometowns;siblings and children all migrate to cities while only elderly parents remained behind to safeguard their old homes with hopes that their children would return someday. Since then, our village house had never undergone renovation and now stands as an aged structure.

The period of challenging living conditions through which I went taught me the

The First "Two-Story" Building in the Village in 1990s

Wood-Fired Earth Stove

importance of diligence and tenacity, helping me appreciate and be grateful for what I have. It is like an altar of old wine worthy of savoring, reminding me not to forget the path I have traveled no matter where I go, nor should I abandon the original mindset that guides me towards my ultimate goals. In that era of material scarcity, we experienced greater spiritual abundance and formed simple yet genuine emotional connections with people, which are invaluable assets in modern society.

The Self-Pressurized Well

Sitting by the Furnace, a Cellphone Served as a Torch for Reading during Power Failure in 2018

Years Spent in the Small City

In the turbulent 1990s, I had the blessing of becoming the favorite of fortune. Through unremitting efforts and persistence, I was admitted to Xinyang Normal University amidst fierce competition among hundreds of thousands of applicants for college education, which marked an important turning point in my life journey. At the university, students lived in dormitories with eight people per room. which was an area of approximately fifteen square meters. Equipped with four bunk beds, the dormitory was crowded but still well-organized and harmonious. The classrooms were spacious and bright, and the campus was beautiful. During my first year at university,we still used "food coupon" to buy meals. As I gazed upon the towering buildings in the city, a mix of surprise and longing filled my eyes, making me feel both humble yet full of infinite possibilities.This spectacular "steel forest" planted seeds of exploration and a yearning for the vast world deep within my heart.

After graduation, with excellent academic records and a passion for education, I was fortunate to become a teacher at my university. The dormitory assigned to me was located in an old seven-story building that appeared shabby and cramped. Originally, it housedtwo people sharing one room witha communal bathroom. There was no kitchen, so everyone had to cook in the communal corridor. When it came time to cook, the corridor would fill with smoke and various food smells mixed together while constant coughing caused by the smell of chili pepper could be heard. Cooking while chatting leisurely—what a vibrant lifestyle!However, the public restrooms were notoriously dirty, making the daily routine a painful ordeal. Despite having tap water and electricity supply, the living environment during this period was even worse than that in rural areas due to insufficient private space and cramped public areas.

After getting married, I was assigned a small one-bedroom apartment with an independent bathroom and kitchen. Despite its small size, I enjoyed the marginal welfare of an improved living environment: having private space, a clean bathroom without any odors, and a quiet kitchen. Compared to my rural childhood life, the

The Communal Corridor in the Old Tube-Shaped Dormitory Building

living environment had truly been upgraded, and I no longer struggled for survival. The apartment was equipped with tap water, electric lights, telephone, TV, air conditioning, and other modern home appliances.The convenience of these modern facilities made me deeply appreciate the progress of the era and the improvement in living standards. Gone were the days when I had to rush to fetch clean water from a well as a child; there was also no need to study under dim candlelight anymore. However, this dormitory building did not have an elevator so it still required effort going up and down stairs every day despite providing some physical exercise.

Metropolis Time

In the early autumn of 2007, I set foot in Beijing, a city that blends ancient traditions with modern aspirations. With a thirst for knowledge and dreams for the future, I embarked on my journey as a doctoral student at a prestigious university here. Beijing, the heart of our country featuring its unique historical heritage and vibrant contemporary atmosphere, warmly embraced me—a student hailing from Xinyang City. Comparatively speaking to Xinyang City, Beijing's housing exuded refinement and advanced architectural style as well as better interior decoration; however, it also came with higher prices.Behind the city's prosperity lay an unspoken acknowledgement of both motivation for success and practical consideration regarding living costs.

After completing my doctoral study, I chose to work in Beijing, a place where dreams and challenges coexist. However, the expensive housing prices were like insurmountable mountains, and I had to lease a place for a transitional period. Initially, I lived in an old residential community with a shabby and confined dwelling furnished with outdated furniture, which made me feel uneasy as if I were living in someone else's house. Being without a stable dwelling, I had experienced frequent relocations, which made me think that staying in Beijing was the wrong choice. At that time, my living conditions were even less comfortable than those in Xinyang in some ways; however, I understood that all those struggles were only temporary—a stage that I had to go through in pursuing higher life goals.

Time flies. In 2012, after years of relentless efforts, I finally achieved a long-cherished goal—owning a home in Beijing. This time, it was no longer just a rented "temporary residence" but rather a true sanctuary for me. I purchased an apartment in a high-rise building with elevators in a new neighborhood. The community environment was beautifully adorned with flowers and trees.Small parks, fitness facilities, parking lots, and household services all contributed to creating an inclusive and advanced living environment that enhanced the residents' overall well-being. I dedicated myself to decorating and furnishing the new apartment: every corner revealed traces of meticulous designs.The incorporation of smart home appliances further elevated our living standards by making it more intelligent, convenient and enjoyable. Although the size of the house wasn't particularly spacious, its well-planned layout and modern amenities generated warmth and happiness within this small yet cozy haven.

Fitness Facilities at the Community Square

Public Garden in the Community

Standing in the spacious and bright living room, looking out of the window at the vibrant cityscape, my heart is filled with a blend of emotions. Along the way, I have witnessed personal struggle unfolding and observed significant changes in the living environment bestowed by this era. I remember that when I first arrived in Beijing, feeling perplexed and uneasy while confronting exorbitant housing prices; however, it was these challenges that inspired me to keep moving forward and made me cherish every opportunity for self-improvement and a better life.

The society is progressing each passing day. Every change in life and the living environment exhibits the evidence of our persistent efforts and achievements despite all the struggles. The transformation of living conditions indicates both the epitome of personal stories and marks of the era's development, serving as a record of our growth while reflecting the progress of the society, for which we are both witnesses and creators. All these accomplishments inspire us to unremittingly pursue our dreams for a better future.

Hometown, My Lifelong Treasure

by Yang Yueguan

Country Life Time

My hometown is Yangjiaba Village, Sanchahe Town in Yunnan Province. It is situated in the renowned Luliang Flat Ground, which is considered the foremost plain on the Yunnan-Guizhou Plateau and boasts favorable agricultural conditions with a dry and comfortable climate. My mother gave birth to six children, including three boys and three girls, and I was the fifth one. Before the National Land Reform Campaign in 1950s, my grandfather, great-grandfather and their brothers used to inherit a two-story wooden farmhouse with a courtyard located in Yangjiaba Village. However, during the Campaign period, they were all classified and labeled as "Landlords", resulting in complete deprivation of their properties including lands by the Land Reform Work Team. Their quadrangle courtyard was also taken away and distributed among other farmers; thus our family had no choice but to build a humble rammed-earth cottage behind the original courtyard. During my childhood years until I turned seven years old when I started primary school education, my family of eight people—consisting of my parents, siblings and myself—resided in this cramped adobe dwelling. The cottage was constructed with a wooden structure while its walls were supported by a rammed-earth foundation. It featured two floors; the second floor served as living quarters while pigs were raised on the first floor. During the period of Collective Economy (Collectivization), farmers had to work in local production teams to earn their working points and could only receive grains allocation at year-end distribution time. In early morning after breakfast, my mother would go out for work earning eight work points per day before hurrying home to take care of a large brood of children. Those days were truly hard.

After the Land Reform Campaign, my family purchased two rooms from our original quadrangle courtyard. Family members could now live in both upstairs and downstairs rooms, while keeping the pigs in the rammed-earth cottage. Consequently, this new arrangement offered a slightly more spacious living space for a family of eight people, particularly when elder siblings departed for education or marriage. The dwelling did not have a separate kitchen; instead, the kitchen and living room were combined into one room.

The National Family Planning Policy was not implemented at that time. Each family residing in this quadrangle courtyard had four to five or even seven to eight children. When I look back at those days, although the living conditions were harsh, I consider them as the happiest days of my life: there was no pressure to live, and I played with my friends in the courtyard every day after supper. The courtyard was always filled with a riot of sound. Now my childhood friends have reached the age of grandparents, and most of them have moved to Luliang County to help take care of their grandchildren. I lived in the courtyard from elementary school until I got married. In 1996, after the birth of my eldest daughter, I accompanied my husband to Kunming, the capital city of Yunnan Province. During my children's winter and summer vacations in subsequent years, we would return to our hometown and stay in a newly-built house for a while.

Collapsed Old Cottage

(Photo taken in 2024)

Quadrangle Courtyard in a State of Long-time Disrepair

(Photo taken in 2024)

Only two rooms in the quadrangle courtyard were too crowded to accommodate such a large family. In fact, the per capita housing area of my family was less than one-tenth of the average level in the village. Therefore, in 1997, my father applied to the Land Administration Center in Sanchahe Town and was approved for a piece of land allocation based on household size to build our own house in Yangjiaba Village. We constructed a one-story brick house with a flat roof for drying goods along with a water reservoir. The construction cost was not expensive, only a few tens of thousands of RMB. Wages at that time were very low, and workers were paid only five yuan per day. Local craftsmen were hired for the construction. In the village, craftsmen did not work out a design plan before building a house; instead, they would calculate all necessary measurements based on their experience once informed of the land areas and desired number of rooms.

The new house had four large rooms, four smaller rooms and two halls. The interior was about 4.5 meters high, and there was a stairwell leading to the roof for drying goods. The whole courtyard measured approximately 1,200 square meters. In front of the house was a large vegetable field planted with various fruits and vegetables to provide the family with fresh produce throughout the year. My second eldest

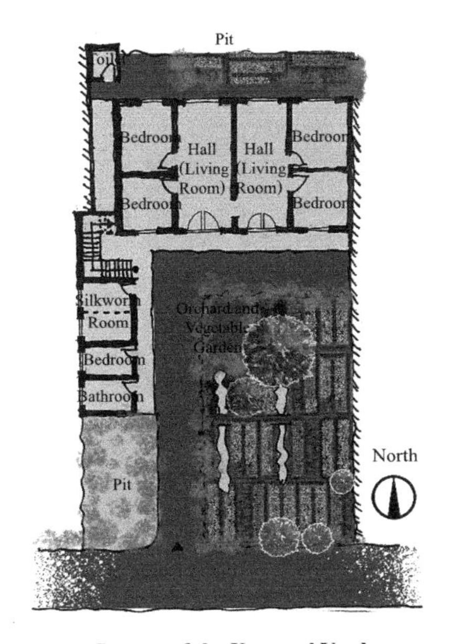

Layout of the Unpaved Yard

Paved Yard of the Old House in Hometown

(Photo taken in 2024)

Inside and Outside the Hall (Living Room)

(Photos taken in 2024)

My Mother and my Eldest Daughter in the Vegetable Yard

(Photo taken in 2000)

brother loved fruits and enjoyed collecting fruit seeds for planting; he used to plant a big apple tree in the yard along with crab-apples, peaches, pomegranates, grapefruits, passion fruits, and grapes. To facilitate watering of the vegetables, a small trench was dug in the vegetable field. In recent years, my mother has become too old to take care of the yard, so it was paved with concrete; only leaving a small patch as a vegetable garden remains now. At that time, there were no walls surrounding each house's yard except for separation walls between neighbors; simple mulberry fences were installed which allowed people an easy glance at activities in their yards. However, later on, each family built high walls and solid gates to close off their dwellings, and I am not sure about the reasons behind this. My parents have been living here with my second eldest brother's family for a long time while engaging in agricultural business. For several years, my second eldest brother and his wife also raised silkworms by constructing shelves in their living room and west room. The family's paddy field was contracted out to an agricultural company for unified greenhouse planting, leaving only a small plot of dry land used for growing grains and vegetables consumed by the family. As my mother was getting old, the small plot of dry land had been taken care of by my second eldest sister and her husband. Every year during the Spring Festival, all the children would be brought back to hometown for a family reunion.

Overlooking a Corner of Luliang Flat Ground

(Photo taken in 2020)

Urban Life Time

After the birth of my eldest daughter in 1996, my family moved to Kunming and resided there for 18 years. Throughout this period, we lived in three different places, all situated in the Guanshang Region of Guandu District. My husband uesd to work for China Railway 16th Bureau Group Corporation Limited, and his company purchased a piece of land nearby to construct a staff community. They offered apartments to the staff members at a discounted price, with the company covering half of the price while we paid for the remaining half ourselves.

Eventually, we acquired our own apartment in Kunming with two bedrooms and one living room, which was priced at 1,700 yuan / ㎡. We resided there for about four years before selling it at a reduced price to compensate for my husband's loss on an investment in a construction project. Since then, we have been renting residences for our family residence.

We rented a three-bedroom apartment in the nearby Tobacco School community. My husband and I used one room, while our two daughters shared another one furnished with an elegantly decorated bunk bed which we called "The Princesses' Bed". Occasionally, my younger brother would come to stay with us and use the third room. Since it was located in the residential community of Tobacco School, the facilities of the community were well-maintained and complete with ample public activity space. There were two sets of large tennis courts, basketball courts and a beautifully landscaped garden. Many children around the same age as my daughters lived in the community, happily running and playing together while we adults chatted in nearby pavilions. We have been living in this neighborhood for a long time, accompanying our two daughters throughout most of their teenage years. In 2014, after my eldest daughter went to college in Beijing and my second daughter enrolled in a high school in Qujing City, we moved to Qujing and rented a two-bedroom apartment near the high school for a monthly rent of 1,200 yuan. Although it was an old community, it had convenient living facilities such as food markets, shopping malls and parks nearby. Moreover, the clothing store where I worked was also within walking distance. Further away from our apartment, we could climb Liaoguo Mountain and do morning exercise on weekends. The only drawback was that some tall buildings in the surrounding community blocked sunlight from entering our apartment; thus reducing the duration of indoor sunshine even more during winter months. Later on, since the community was about to be demolished, we had to move out. Life in cities is similar and always keeps a fast pace. However, when comparing Qujing with Kunming, I would say that due to having more rural areas around Qujing, its living cost is relatively lower and its natural environment better. Nevertheless, I will choose several memorable moments from my time spent in Kunming to illustrate my actual living experience there.

Anti-theft cages were commonly installed in old residential communities in Kunming, and almost every household would be equipped with them. Those iron or aluminum alloy cages were installed outside windows, usually along the streets or on sunny sides, either to prevent thieves from entering or to dry clothes, bacon, and plant flowers. The cages did affect the appearance of the city; moreover, they posed risks of falling objects through the barriers. However, due to a lack of sufficient monitoring systems, the installation of such cages became necessary. We used to live in an apartment on the fourth floor without an anti-theft cage. One morning when I woke up, I was stunned to find that my home was in a mess and my wallet had been emptied! Later on, we found a black palm print on the white wall outside the balcony windowsill, which implied that a thief must have sneaked in from the balcony at night. After this incident, I felt a moment of fear and still do not know how the thief managed to climb onto our fourth floor apartment. Subsequently, we moved to a fifth floor apartment with an installed anti-theft cage and since then we have never experienced thefts.

Anti-Theft Cages

When I lived in Kunming, there was a famous village street near my home called "Guangai Street" by locals, and the market operated every Saturday from 6 am to 7 pm. The location of "Guangai Street" belonged to a transitional zone during the urbanization stage of Kunming, where some peasants still owned their lands and lived as rural residents within the urbanized areas. Every Saturday, vendors from surrounding areas would utilize open spaces and rural roads to set up various tents and organize a bustling flea market. The market not only sold agricultural products but also offered daily necessities, farm tools, building materials, food and beverages, birds, fishes, insects, flowers, antique treasures, herbal remedies as well as tooth extraction services and open-air barber stalls. The market boasted an extremely lively atmosphere reminiscent of New Year celebrations.

Vibrant Guangai Street

(Photos sources: Kunming Headlines and Colorful Dragon Community of the Kunming Information Port)

During a certain period of time, my family became accustomed to visiting "Guangai Street" every Saturday without any specific purchase intention; instead, we would enjoy the vibrant atmosphere there and try some delicious fried potatoes. However, we always ended up bringing back a lot of fresh fruits or local products. During this period, the old "Guangai Street" was closed several times for rectification in order to address issues such as overcrowding, disorderliness and village renovation. Nevertheless, nearby farmers would sell their products in other places and gradually formed a slightly smaller-scale version of "Guangai Street", which continued to provide convenience and serve as a nice marketplace for nearby residents.

In addition to watching the art of bargaining in the market, it was also fun to observe aircraft taking off and landing from a close distance. "Guangai Street" was located near Wujiaba Airport, which was still operational at that time. Every now and then, you could see planes taking off and landing from the marketplace,

A Plane Flying Overhead Across Guangai Street

(Photos source: Wujiaba in the River of Time—the 2nd Oriental Brocade Photograph Exhibition. Photographers:Weiming Ruan and Jianming Liu)

creating a loud buzz in your ears. Looking at old photos of planes above the streets on the Internet now gives me a magical feeling. However, this scene is now part of history as Wujiaba Airport ceased operations in 2012 after the construction of the Changshui Airport, and the latter has become much more renowned among citizens. After we left Kunming, we later learned that "Guangai Street" was permanently closed at the end of September 2017. The Wujiaba Area and Guangai Area are now

Farewell to Wujiaba

(Photos source: Wujiaba in the River of Time—the 2nd Oriental Brocade Photograph Exhibition. Photographers: Ming Liu and Jianhua Liu)

planned for development into regional centers with numerous large commercial and residential facilities. In this rapidly urbanizing era when urban and rural integration is taking place, the hustle and bustle of "Guangai Street" can no longer keep up with the pace of urban development; thus lagging behind in history.

Working Outside

In 2021, after my two daughters had started their careers in northern China, the main goal of my life was essentially achieved. Due to some changes in my family situation, I made the decision to seek employment elsewhere. After being introduced by my relatives, I went to Ningbo City (a developed city of Zhejiang Province located in southeast China) and found a job at a factory in Huangjiabu Town, Yuyao County. I rented a farmhouse in the nearby village for residence. Many migrant workers who lived in Huangjiabu Town came from Luliang Flat Ground because they had heard that there were many factories offering higher wages in Ningbo; thus they were introduced one by one to work here.

The living conditions in Ningbo were simple yet tough: the rented place was a single room with low ceilings and a tiled roof, measuring about 10 square meters in size, and it costed me 300 yuan per month. The room was dimly lit, with a brick cooktop for cooking, a bed separated by a curtain from the stove, and a shared bathroom in the courtyard. As I went to Ningbo with a sole purpose of earning money, I paid little attention to living conditions. The climate of Ningbo was distinctively different from that of Yunnan Province where people could enjoy spring-like seasons all year around. The weather in Ningbo tended to be colder and consistently rainy—its lengthy rainy season made outdoor activities quite inconvenient. In summer, it would become extremely hot, sometimes even exceeding temperature above 40℃. I spent long hours working daily at the factory without much personal time. The only bit of fun was digging shepherd's purse or bamboo shoots in the mountains with fellow villagers on my days off.

In 2023, my second daughter gave birth to a baby in Beijing, so I went to help her for over half a year. Her family rented a courtyard with a two-bedroom and one living room house, and I lived with them. My eldest daughter also rented a place in Beijing and often visited us on weekends when she did not have work, staying with us at night. It was difficult for me to adapt to the climate of Beijing. On the very first night I arrived in April, I experienced a dust storm that I had never seen before. As soon as I got off the high-speed train, my eyes immediately felt hazy. During winter, plants along the roadsides were withered; there was no green scenery outdoors which made me feel quite depressed. Previously, I used to walk every day and do outdoor exercises as part of my routine; however, during Bejing's winter season, I had to stay at home for long hours which was unhealthy for people especially kids.

Transformation of Hometown

In the Spring Festival of 2024, I returned to my hometown and noticed significant changes, particularly in the village environment. Over the past two or three years, cement roads and street lights have been constructed in front of each household's courtyard, greatly improving traffic conditions. Every day, people drive electric tricycles or vans to sell a wide variety of goods to villagers. If you need any thing, simply greet the vendors with a hello and they will bring it right to your doorstep; there is no need to go to the town center anymore. Additionally, shared electric vehicles are available for rent throughout the village, and taxis can even be found here now, making transportation to town much more convenient than before. The nearby Qinghe Community of Sanchahe Town has also developed several scenic spots that attract tourists who take photos and share them on social media platforms. These spots are adorned with antique items from the village hanging on

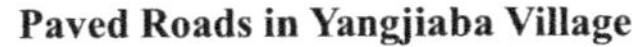

Paved Roads in Yangjiaba Village

Model Village of Qinghe Community for Rural Revitalization

(Photo source: Public Account of National Rural Revitalization Research Institute of China Agricultural University)

walls, and flowers are planted everywhere. Even local residents enjoy visiting these spots as well. Overall, rural environment has dramatically improved.

Most families in the countryside have only elderly people living in the village, and so does my family. When the children of my second eldest brother grew up and left hometown to study elsewhere, their parents also left to work in other places. As a result, only my mother resides in our hometown courtyard. Concerned about the lack of filial care for our elder mother, my youngest brother installed a camera in the yard. With it, if there is any accident, nearby relatives can be immediately notified and help take care of her.

Reflectionon Residence

I have been living in rented places all these years, so I haven't cared about decorating these temporary dwellings. During my stay in Kunming, I embroidered several large cross-stitch works that I had imagined hanging up in my future home, but they were never displayed. After relocating several times with my family, I stored them at the bottom of my box and lost interest. At the age of 50, it is my wish to have a dwelling of my own—a permanent place to spend my twilight years. If given the choice to settle down, I would prefer Qujing City. Practically speaking, the pressure of buying an apartment in Kunming is somewhat too overwhelming for me; moreover, I have spent a long time living in Qujing City and am familiar with its environment. On weekends, I can ride an electric bike to the nearby mountains for hiking; starting from July, I will be able to go into the mountains to pick wild mushrooms. Winters in Qujing are not very cold, allowing people to take walks in the parks. If my daughters return to Kunming in future years, the intercity trains between the two cities will reduce travel time by just one more hour. It would satisfy me if we could meet every two weeks or once a month.

Note: The term "Jiaxiang Bao", literally translated as "Hometowns are Treasure", is a dialect of Yunnan Province. It signifies that Yunnan people love their hometowns and consider them the most valuable places, no matter where they go. Even if we leave our hometowns, we will eventually find a way to return.

Dwellings

by Zhai Jiahui

For modern individuals who have spent considerable time wandering through cities, a residence means a place to settle down that offers the possibility of rejuvenating both their minds and bodies. It has become increasingly challenging for young people who make a living in large cities to stay in one place for an extended period, resulting in even shorter memories of their temporary dwellings. As a young man who left the hometown not long ago, I still cling onto numerous memories associated with it. At present, being twenty years old, I have borne

An Empty House in a Village Labeled as "No Residence"

(Photo taken on May 5th, 2024)

Abandoned Mud Brick Houses in the Village

(Photo taken on May 5th, 2024)

witness significant transformation occurring within my hometown over the past two decades. The process of urbanization is unstoppable, and the countryside has also achieved remarkable development; however, numerous rural houses stand as vacant shells due to the increasing population leaving during the process of urbanization.

In the summer of 2004, I was born in Zhaili Village, Yuecun Town, Hualong District, Puyang City, Henan Province of China. It has been twenty years since I lived there. It seems that the village has not changed much; yet,upon reflecting on my memories, I suddenly realize that the village has undergone a great transformation.

Located on the outskirts of the City, our village is at a medium level of development compared to other villages in the entire district. Over the years, there have been few changes in terms of type and appearance of village houses—they are all tall and thick brick-built structures. There are even occasional old houses with grey brick pillars and mud brick walls; however, almost none of them are currently inhabited.

Our home has remained the same for twenty years: two main halls with peaked roofs, two flat-roofed western rooms, a bathroom under the attached stairs of the western rooms, two small sheds and a toilet on the eastern side of the yard. Our yard is quite spacious; my grandma always plants vegetables for different seasons on a piece of land in the yard. She grows onions, garlic seedlings, lettuce, loofah, cucumber, pepper, ginger... Whenever you want to eat some, just go ahead and pick them at your will. Along with vegetable fields are Chinese toon tree, elm, walnut tree, persimmon tree, etc., which tenderly shade the yard and generously provide

Hometown Courtyard

(Left: Photo Taken on February 5th, 2024; Right: Photo Taken on May 5th, 2024)

nuts and fruits in different seasons. When I was a child, I raised chickens in the east shed of my home. The fluffy yellow chicks grew up little by little and became brown-feathered hens or roosters with fiery combs and gorgeous tail feathers. Their crow echoed through the yard adding vitality to the entire place.

When I was young, the harvest seasons of summer wheat and autumn corn were a god-given happy time for me throughout they ear. During the summer, the tawny grains of wheat would be brought home by cart and spread out in the yard on a thick layer of oiled paper. Grandpa would walk barefoot up and down to evenly spread the grains to get them dried. As a kid, I liked imitating adults' behavior but often did so too quickly and anxiously, kicking off a lot of wheat from the oiled paper, resulting in grandpa's loud stop. He would then pick up each grain one by one and place them back onto the oiled paper. In autumn, Dad drove a tractor to transport corn from the fields and stacked them into a hill in the yard. My brother and I always enjoyed climbing to the top of the corn hill, sitting there while watching our family busily engaged in farming chores. Once we had enough playtime, we helped grandma and mom peel off husks from the corn; however, it didn't take long for us to become fascinated by small white worms found inside the corn and start playing with them instead. During these busy harvest seasons, not only did our family have an atmosphere of good harvest, but all streets were also paved with either wheat or corns; thus creating an entire village bustling with both busyness and joyfulness.

The village has left me with many good memories of my childhood, and I cherished the bustling harvest seasons that are now almost impossible for children to experience. Nowadays, most villagers lease out their lands to collect rental revenues, so they no longer have to work hard in the fields; as a result, the busy yet joyful harvest atmosphere has also disappeared, and in the village children probably never get to experience the happiness of my childhood.

The overall appearance of the village changed dramatically after cement roads were built when I was in primary school. Previously, there were only dirt roads, which would become muddy on rainy days. With the paving of cement roads came significant improvement in enhancing the environment of the village. Additionally, numerous infrastructures were renovated by the Village Committee, bringing remarkable changes throughout the entire village.

There is a temple in our village called Duntai Temple, which enshrines the Duntai

Village Streets

(Photos taken on May 5th , 2024)

Lord. On every important holiday of the year,such as the Spring Festival and Lantern Festival, every family would visits the Duntai Temple, burning incense and kowtowing, expecting the blessings of peace and success for their families. When I was very young, Duntai Temple was a tiny brick house furnished with only a portrait and an old wooden table with an incense basket placed on it, allowing only three or four people to stand at a time to offer incense and pray for good luck. When I was in primary school, the Village Committee raised money from the villagers to rebuild Duntai Temple into a small but serious temple that also functions as an ancestral hall to record the history of the village. The new temple features red walls and gray tiles. Inside, there is a large statue of Duntai Lord. The walls are

Duntai Temple

(Photo taken on May 5th, 2024)

painted with images of heavenly beings such as Nuwa and Fuxi recorded in ancient mythology. Surrounding the temple are stone tablets inscribed with information about ancestors who first moved to the village, the history of our village, and genealogy of past generations. I have always believed that Duntai Temple is the root of our village, representing the core of our culture. Villagers with the same surnames gather here with a clear sense that we are one family sharing common ancestry, bloodline and history.

In recent years, the village has been connected to tap water and natural gas, renovated public toilets, installed solar panels, and added numerous pieces of physical exercise equipment on the streets. As a result, the overall level of village infrastructure has improved. Additionally, more supermarkets and delivery service stations have been built, making living in the village much more convenient. However, it is a fact that fewer people choose to stay, especially young people; consequently leaving only the elderly behind. Most of the remaining young individuals do not opt to reside in their old village houses; instead, they prefer purchasing apartments in newly developed residential communities located to the south of the village.

Real Estate Development of "New Village Community"

(Photo taken on May 5th, 2024)

In 2022, Puyang East High-Speed Railway Station was put into operation. Even before the construction of the station, many real estate developers had initiated investment and property development in outskirt regions around the city. Many villagers located to the west of our village had their lands expropriated with compensation earlier due to their location advantage (more proximity to urban areas). Subsequently, villagers received distributive apartments at the cost of losing their traditional courtyards and the lands that they had lived on for generations. These newly developed communities have great appeal to villagers due to the availability of better education resources for children and their inevitable involvement in modern urbanization. Therefor, many villagers eventually chose to settle down in these new communities. Before the construction of the station, many villagers held a neutral attitude towards purchasing apartments in urban areas. However, after its construction, more villagers preferred living in these new communities with improved residential facilities favoring its convenient transportation and better development prospects despite having higher mortgages because of the almost doubled housing prices in surrounding areas.

Puyang East High-Speed Railway Station
(Photo taken on February 2nd , 2024)

Residential Buildings near the Puyang East High-Speed Railway Station
(Photo taken on February 2nd, 2024)

More and more high-rise buildings have gradually taken away the population and prosperity from our village, making it less lively and enjoyable than before. I no longer feel the same connection with the lands as I used to, especially during the busy harvest seasons.

A dwelling is more than just a place to live; it represents not only a housing facility but also the overall environment in which an individual spends their lifetime. Over these years, dwellings across China have slowly changed along with the transformation of modern life. Upon reflecting on my long-time residence in my home village, I definitely sense that the rural areas are being gradually swallowed up by cities. Although living conditions in rural areas have significantly improved, young people are moving to cities, causing today's countryside to slowly become an empty shell.

Since I left for high school, I have gradually moved further away from home and spent several years living in temporary dwellings. During these drifting years, I have come to realize that I do not enjoy living in a restless place. Instead, I prefer to have a peaceful abode of my own, which does not necessarily need to be spacious but should provide rest for both my body and soul. However, achieving this wish is quite challenging in reality due to the expensive housing prices that impose immense pressure on young people like me throughout their lives. On the other hand, residing in the countryside can hardly provide young people with opportunities for promising development. Being homeless in the city while feeling lost in the countryside is my dilemma and also the hidden pain experienced by the contemporary youth.

Cities embody social development while countrysides serve as the roots of our culture. If we could strike a balance between the advantages of both settings by creating more career prospects in rural areas and facilitating more young people's return to their hometowns, it will undoubtedly contribute to rural revitalization, youth development and cultural inheritance within our country. With nationwide implementation of social modernization, I hope that this day can become a reality soon!

From cramped living to livable life

by Zhang Ting

In Du Fu's *Lyrics of the Thatched Cottage Broken by Autumn Gale* (*translator's note: Du Fu, born in 712 and died in 770, is a star poet of the Tang Dynasty and often regarded as one of the greatest in the history of Chinese literature.*), there is a famous sentence as follows:

Tens of thousands of resplendent mansions,
How can I transform them into grand havens?
Sheltering under their domes,
Bringing blissful homes and broad smiles to people with little means.

It has revealed people's long-lasting dream of living and working in peace and contentment. From spring to autumn, flowers bloom and fall; rivers flow steadily from west to east, while leaves sprout and fall one after another. 75 years have passed since the establishment of the current Chinese regime, during which we have experienced significant residential transformation from thatched cottages and small mud huts to high-rise buildings in modern communities; from barely being sheltered from wind and rain to a beautiful and comfortable environment. Time flies as my motherland continues moving forward each passing day, and we no longer live in the same way as before.

From Mud Hut to Brick House with A Tiled Roof

I was born in a small village in the 1980s. My parents were simple farmers. The countryside was my childhood home, consisting a modest yard and several poorly constructed mud huts.

By pushing aside the wooden hedge, one could see a separate area where my parents kept their poultry. This was followed by three huts with mud-thatched walls which were faintly cracked and weathered from years of wind and frost. The wooden window frames were covered with newspaper, and only two or three had pieces of scavenged glass for light. Upon entering through the door, there was a kitchen dividing the living room into two sections. The inner section served as a bedroom where an earth *Kang* (*translator's note: traditional brick/earth bed that can be heated for winter in northern China*) was installed. Mats covered the top of the *Kang*, while its bottom featured tunnels connected to a chimney for heating purposes using firewood. Huge bamboo mats spread across the *Kang* along with piles of bedclothes—this was my home; it may sound shabby now, but it truly provided haven during my childhood.

In 1994, we moved from the mud huts to a brick house with a tiled roof. The courtyard was sealed off with bricks and cement for the gate and fence, which made me feel much more secure. Upon entering through a big red iron gate, one could see a spacious yard. To the left and back of the brick house were four small sheds: one for storing farm equipment, one for raising poultry, and one for stacking firewood. On the front left side, we planted a small apricot tree with its trunk interwoven with branches, and its surrounding clearing was used as a small vegetable garden where we grew some vegetables for daily consumption. Further ahead was the brick house equipped with dazzling glass windows. It boasted five spacious and bright rooms, a cement *Kang*, and a complete set of furniture that was truly elegant and comfortable at that time.

New Brick House with a Tiled Roof

Spacious Courtyard

During those days, most phones were landlines, and TV sets were not popular at all. In fact, only a few families could afford an old-style TV set. After school, kids often gathered happily around the TV to watch episodes of ***Journey to the West*** together.Although the TV was black and white in display, children enjoyed it very much.

From a Young Migrant Worker to Starting a Family

In 1999, when I was only 17 years old, I left the countryside for the first time and came to the city to explore the outside world. I never expected that the outside world would be so wonderful. Cars and bicycles were coming and going on the wide tar roads, while shops and high-rise buildings were scattered on both sides. The scenes of the city fascinated me.

At that time, I was working in a hotel and living in an ordinary staff dormitory—with rough concrete floors, bunk beds, six people sharing one room which felt slightly crowded. At the end of the corridor was a toilet where I used tap water for the first time in my life. I had to admit that the living conditions in the city were much better than those in rural areas. I worked and lived there for over two years.

When I got married, my living situation changed from residing in a dormitory to renting a house. Positioned within a small courtyard, the rental house was oriented towards the north and had two stories where each floor accommodated one family. My home consisted of only two rooms without an exclusive living room, and our furnishings were limited to just one big bed and cabinet. Despite being merely four walls lacking housing amenities, life there remained bustling but full of promise. The kitchen was located at the far end separated by a storage shelf; meanwhile, we shared a bathroom installed in the downstairs corner with another family. Due to its inconvenient location far away from the city center which made transportation difficult for us, we relocated after enrolling our child into primary school.

The new dwelling refreshed my understanding of residential buildings. Rough concrete gave way to marble floors, peeling whitewashed walls were replaced with wallpaper, and the spacious living room was furnished with a tidy sofa. Living conditions significantly improved thanks to access to tap water and natural gas supply, floor heating system, and an independent bathroom. In summer, an electric fan cooled us with refreshing breezes and a refrigerator kept watermelons chilled; in winter, soft and fluffy cotton quilts wrapped us in cozy warmth and comfort.A washing machine spared us from hand-washing clothes during cold winter seasons. The overall living experience had been greatly enhanced.

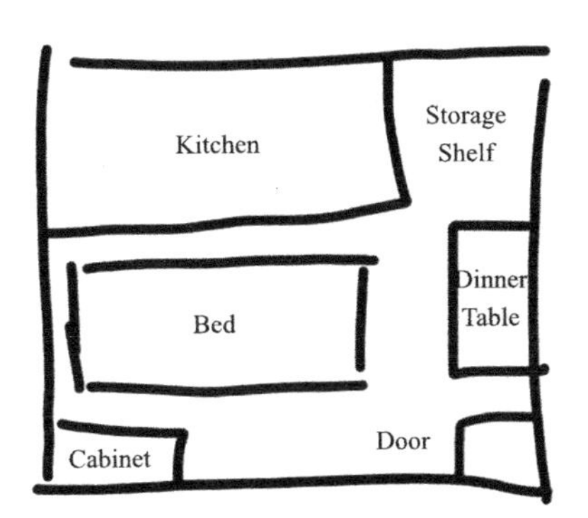

A Sketch of the Rental House Layout

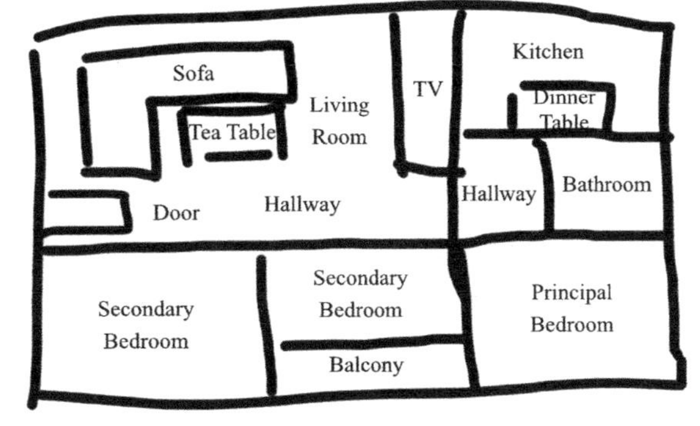

A Sketch of the Modern Apartment Layout

Life in Xingjiang Province

In 2016, our child enrolled in junior high school.Due to work reasons, the family relocated to Xinjiang Province. Initially, I believed that the western region must be far less prosperous than the central provinces with vast deserted lands and sparse population. However, I was completely amazed when exploring its picturesque scenery, abundant resources and talented individuals. There were endless wildernesses, golden sunflower fields, and ancient dragon roads on Pamir

Golden Sunflowers

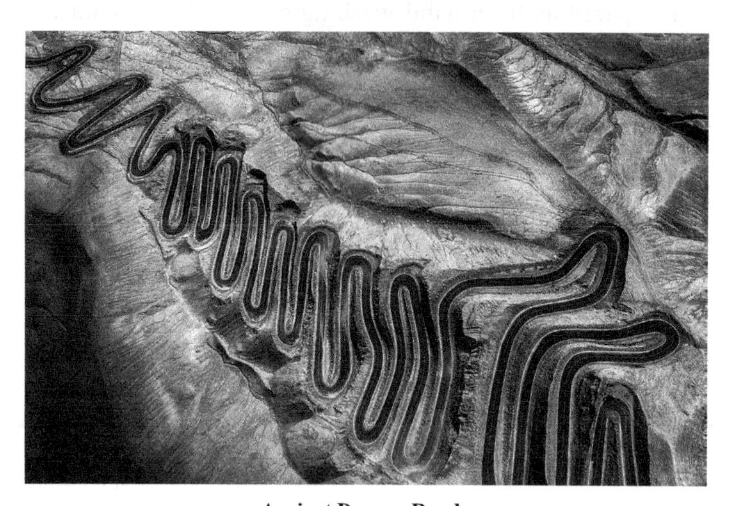

Ancient Dragon Roads

Plateau resembling a dragon hovering around mountains. Reflecting on the ancient Silk Road over a thousand years ago when caravans passed through and camels transported silk and tea along it to West Asia and Europe, I exclaimed at what long and challenging journeys they had trodden upon.

After spending our second year in Xinjiang, wc purchased our first apartment. Constructed with reinforced concrete, it is solid and secure while being particularly comfortable after undergoing custom renovations. Smart home appliances are

My Sweet Home

Parking Shortage in the Community

contemporary devices that provide incredibly convenient and comfortable use. The residential community features a beautiful environment. Additionally, there is a kindergarten within the community along with nearby schools, which offers great convenience for families with children.

From the original "cramped home" to the current "comfortable home", our living standards have significant improved. Nevertheless, there are still some shortcomings. For example, nowadays almost every family owns a car, but there is no underground parking lot in the community; therefore, the available parking space becomes so limited that people often struggle to find a spot.

In the past 40 years of my life, from rural to urban areas, and then to modern metropolises; from mud huts to brick houses, and eventually to today's high-rise buildings; from wooden furniture to household electrical appliances, and now to smart home appliances. All these revolutionary changes symbolize our national progress. The level of comfort experienced in residential areas not only indicates the development of cities but also represents that of the entire country. Our motherland is prospering continuously, and its people are leading increasingly promising lives.

Memories of Hometown Roads

by Zhou Chenhao

I often heard the elders say that it used to take a whole day to reach Jianyang City (located in Sichuan Province, southwest of China) from our village. At that time, there were few cars, and people were poor. Villagers had to walk on unpaved roads to get to the city on foot. People could hardly afford rain boots, and some of them even walked barefoot. If it rained, they had to struggle through puddles and splashed mud, with their feet unable to be pulled out or even getting hurt with scratches. Whenever I heard these stories, I would feel sorry for how difficult the situation was at that time.After all, in the first twenty years of my life, I have witnessed the transformation of roads in my hometown from mud roads to asphalt-paved ones.

Memory of Muddy Roads

I still vividly remember the first time I greeted my Mom and Dad upon their arriving home when I was three years old. After being informed of their imminent return, I immediately ran out along the road they would pass by. The dirt road was filled with little mounds and potholes everywhere, causing me to repeatedly fall and get up while running as quickly as possible. When I finally met my parents, tears burst from my eyes. Children in our village were all like this: brimming with excitement when parents returned home for the New Year celebrations because we children had been left behind in the village all year long. While boys would run along muddy roads to welcome their parents back, girls would eagerly stand at the gates until catching a glimpse of their beloved figures approaching before rushing towards them.

I was very mischievous during the two or three years in kindergarten. At that time,

The Old Muddy Road

my parents worked as migrant workers in Guangzhou City, far away from home, and my Granny took on the responsibility of taking care of me, including sending me to kindergarten everyday. The road from my home to the commune kindergarten was filled with mud. If it did not rain, Granny would lead me and walk to school together. Sometimes I was very stubborn and cried to be carried on her back, but she didn't mind such a small trouble. However, if it rained, Granny had to carry me on her back because the road was muddy and slippery with dirty water. Granny had to be very careful; she always bent down and followed the footprints left by those who came before us. Sometimes it was quite easy for her to slip since there were no stable footholds available. Granny had no choice but to walk straight through other villagers' farmland fields because there was not much water left in the fields when crops grew, making walking easier. Almost everyone walked this way, creating a path as they went along. Although this behavior was wrong, villagers had no option but to try their best not to step on the crops while passing by them. Regardless of whether it was sunny or rainy outside, Granny always took me along the muddy road to school.

I remember several boys and girls who were a few years older than me in our village always played with me and took care of me since I was three years old. Therefore, when I enrolled in primary school, Granny no longer sent me to school; instead, these elder brothers and sisters would accompany me every day. Being the youngest one, I received tender attention from them. On rainy days, we would put on our rain boots and each bring a pair of spare shoes. The muddy road would make the rain boots and clothes dirty and unrecognizable. Sometimes the road was really difficult to walk on, especially going downhill: even if we held hands tightly and stepped extremely carefully one after another, we would still slip again and again. Due to children's nature, sometimes we played along the way by stepping into the mud puddles which splashed mud all over our clothes, boots and everywhere else. Luckily, there was a pond on the side of the road near the school where we could wash our rain boots. As we were afraid of falling into the pond while washing them one by one alone, other kids would pull our clothes from behind.

During that time whenever it rained, I often slipped and cried on my way to school. I kept wondering why we had to traverse such a muddy road. It wasn't until I reached Grade Three that we no longer had to walk on this filthy path.

Memory of the Cement Pavement Road

One day, trucks brought truckloads of rubble while a construction team spread it evenly on the dirt road surface. Since one part of our usual route to school used to be a narrow pathway, this particular section underwent enlargement for reconstruction. The gravel consisted broken bricks and pebbles which caused immediate foot pain for walking even when wearing shoes. However, its advantage became evident as it was no longer slippery on rainy days. After waiting until the rubble had been firmly embedded, the construction team returned to continue their work. Villagers were very excited and watched them nearby. I was also thrilled since I understood that they were about to build a really nice road. The construction team repeated the steps of installing plywood, embedding steel bars, and pouring concrete. They only completed one section of the road in a given day because it took some time for it to be completely shaped. After a few days, the road was fully constructed. Eventually, we could easily go to school even on rainy days.

When the road was newly completed, there wasn't much transportation. Therefore, skateboarding became popular among kids on the rural roads. Dad bought one for me at my earnest request. However, I was still too small and inevitably fell repeatedly. But I remained happy to play nonetheless. Unfortunately, one day I had a bad fall which caused much hurt and pain as my knees rubbed against the road and scraped off the skin. Tears burst uncontrollably from my eyes. Upon seeing this, Dad became angry and broke the skateboard in frustration. Since then, I had never skated again. Gradually, there had been an increase in traffic on the road, especially motorcycles; therefore, children were no longer able to play skateboarding on it.

Although the road leading to school had become convenient, the road to the county was still challenging. As far as I can remember, the road used to be a highway from my home to the county, with a shuttle bus commuting from the town center to the county every day. However, each time I traveled to the county proved painful because I suffered from severe carsickness. I remember that when I was four years old, my grandparents' health deteriorated, so my father returned from Guangzhou City to help my aunt do business in the county. During each winter and summer vacation, Dad would take Granny and me with him to stay in the county. Whether inside or outside of vehicles, the exhaust gas smell was really disgusting. Every time I got on a bus, I had to cuddle up and rest on Granny's legs. The highway leading to the county was in bad condition as many places had been squeezed or distorted by heavy loads of large trucks. The original one-and-a-half-hour drive painfully extended into a two-hour long journey. Moreover, traffic jams often occurred on this deformed road. Sleeping used to somewhat alleviate carsickness before it became ineffective, but eventually it would inevitably force me to miserably vomit every time.

On the night of my grandpa's passing away, I urgently called Dad and Aunt, asking them to hurry home while my grandpa was still holding on. From the moment I made the call until Grandpa took his last breath, it felt like an endless wait. However, Dad and Aunt failed to get home in time to catch a final glimpse. At that moment, I sadly realized how time-consuming the highway from the county to my home was, despite being just over 20 kilometers in distance.

Due to heavy traffic, large trucks frequently took a shortcut by passing through our village roads, resulting in significant wear and tear on the road surfaces. Along with some other factors, these village roads became cracked after a few years of use with some areas even dented. It was not until my first year in the middle school that I heard about new roads being built to replace the original ones in the village. This change was significant as it also involved renovating the highway leading to the county.

Memory of Asphalt Roads

Drilling machines continued to drill into the road, and after a few days, the entire road had been shattered into gravel mixed with steel bars. Once the gravel was cleared, the earth foundation was exposed once again. On that day, the construction team clamped boards on both sides of the road while one driver poured asphalt and another followed by driving a road roller to compact it. Villagers watched with interest and excitement, especially a group of kids like me who trailed behind the road roller at a distance. The fresh asphalt remained hot, black and sticky, but that's what made it so fascinating for me.

After the successful construction of the asphalt road, conditions in the village gradually began to improve. My dad returned home from the county and started his chive planting business on leased farmlands. Other villagers followed suit and also got involved in vegetable cultivation. With collective efforts, they reached out to buyers to sell their products. As a result, the living standards of each household improved. Thanks to a promotion campaign of national policy, most of the production teams that were previously under commune authority began constructing residential facilities for distinct village communities. Villagers had the privilege of owning apartments in these newly built facilities at the cost of giving up their old house land. The size and number of the allocated apartments were calculated

The Asphalt Road Under Construction

The Asphalt Road

based on the land areas of villagers' old houses. Some weal thier villagers from nearby village also invested in family farm business. All roads connecting these villages were well-constructed, attracting many people for on-site visits. More and more villagers enjoyed leisurely strolling through the streets. Sometimes after school, I would see teachers from our commune school passing by my home and warmly greeting me. After supper, Granny and other elder ladies from our village would take walks along the roads and casually chat together or visit nearby villages for fun. I would ride my bike with other kids to enjoy moments of friendship and joyous times together. Since new roads have been constructed, I no longer have to walk on muddy paths for school like before. Occasionally, while taking a shuttle bus and passing by the fork road, I might catch a glimpse of the old road covered in overgrown wild grasses.

I passed the entrance examination and was admitted to the best high school in our county. On the day of registration, I arrived at the bus station early. The shuttle buses had all been replaced with electric-powered ones by then, and there was no longer a pungent smell of fuel that used to linger around them. Having previously taken shuttle buses during every winter and summer vacation before this change occurred, despite their different models and shapes now, sitting on them no longer caused any discomfort for me. In contrast to previous times when county buses needed to take a detour route, it was my first experience traveling to the county on the newly renovated highway. Sitting comfortably, I didn't experience any nausea as it was a smooth ride that only took me 40 minutes.

In my three years of high school, I could not remember how many buses I had taken or how many times I had walked on the asphalt road from the county to my village. It remains firm and solid today, unaffected by the numerous large buses or trucks that have passed over it.

Moving forward for a Promising Future

Looking back at the muddy roads of my childhood, every footprint on them is imprinted with Granny's love. Every tumble contributes to my growth experience, and each time of fellow partners' help embodies their kindness and friendship. During my childhood, the old road once facilitated villagers' travel more easily, but years of wear and tear made it difficult to walk through. By replacing the deteriorating road with a new asphalt one, traffic has become much more convenient. The new road can bear heavier loads and successfully attract people to return home for business. Convenient transportation promises a prosperous future. From mud to cement and then to asphalt, the transformation of roads in my hometown holds our sentimental memories; it is also a microcosm of the changes in people's lives.

Just like the significant changes in our village, who could have imagined that Tianfu International Airport would actually be constructed in our county? This area has been designated as the Eastern New District and has continuously influenced economic development in its surrounding area since the construction of the airport. The construction of any road always implies better lives for people.

My Residentid Impression

by Zhu Lixin

I was born in 1949, the same year that marked the establishment of the People's Republic of China. Since then, I have experienced numerous relocations to different residences. During my childhood, I accompanied my parents as we moved four times within bungalow areas, where there were dirt lanes lined with rows of bungalows. Typically, each row consisted of ten rooms measuring approximately 10 square meters each. In 1960, our family of five—including my father, mother, younger sister, younger brother and me—resided in two such rooms. During that time, locals used "*Kang*" beds (*translator's note: traditional brick or earth beds that can be heated in winter in northern China*) rather than regular beds. These *Kangs* were usually built along the lateral wall of a bungalow and extended from the front window to the back wall; this type was referred to as "*Shun Shi Kang*", meaning "Lateral *Kang*".

Being small kids, my younger sister and brother slept on the big *Kang* with my parents. When I was over ten years old, my father nailed a small plank bed for me and placed it against the wall opposite to the *Kang*. In the middle of the open space, which we might call "living-room area", we carried out all living activities such as receiving guests, cooking and dining.My sister and I even studied there,doing our homework on the dinner table.

There was no kitchen in each house; instead, we placed the briquettes and stoves in the yard. When it rained, we had to carry the stove under eaves for cooking; however, heavy rain or strong wind might extinguish the stove fire. Our family maintained such living conditions until the mid- 1990s.

I graduated from the junior high school in 1966. Due to the "Cultural Revolution",

I was transferred to the countryside in the suburbs of Beijing by the end of 1968 as part of the national campaign "Relocating Youngsters to Rural Areas". I settled down and resided with a local peasant family. At that time,the suburbs of Beijing were also in an impoverished situation. The villagers typically lived in three adjacent rooms facing south, known as "Tiger-Head House". Among these rooms, the middle one was brighter with a gate opening towards the yard,while the other two were darker. The middle room resembled a tiger's mouth, and each of the east and west rooms had front windows resembling tiger eyes.

The middle room was equipped with a stove near the west wall for cooking and also warmed the adjacent western bedroom by heating the *Kang*. A water tank was placed next to the "bag wall" (*translator's note: similar to Kang, a kind of inter-layer brick wall used for heating purpose*). A wooden board was set up against the east wall to hold a chopping board for cooking and provided storage space underneath for firewood, sundries and foodstuff.

The east and west rooms served as bedrooms. According to Chinese convention, the east orientation is regarded as "upper" side. Hence, the east room was installed with a lateral *Kang* along with a stove under the south window, supplying hot water for all the family needs in cold seasons. Simultaneously, its heating channel warmed up the eastern bedroom. In the western bedroom, a *Kang* under the south window was connected to a cooking stove in the middle room through heating channels and also provided heating in the middle room. This type of *Kang* was called "Front *Kang*". These were the basic structures and functions of the three rooms.

A more prosperous family would construct a five-room house, which meant adding one more room on each of the left (east) and right (west) sides based on the three-room type (also known as "Tiger and Leopard Head" by the locals). Generally, these two attached rooms were relatively small, and they looked like two ears hanging on both sides of the main middle structure. Therefore, they were also called "Ear Rooms", with the east one referred to as the East Ear Room and the west one as the West Ear Room.

For wealthier families, they directly constructed houses consisting of five equally-sized rooms. These houses represented the highest standards for rural dwellings during that time and were referred to as "Five Rooms Sharing One Ridge." Affluent families adorned their roofs with tiles, while less fortunate households used plaster stalks instead. Even poorer families resorted to using either "Slippery Straw Mud" (a mixture of yellow mud and broken straws about half a foot in length to increase tensile strength) or "Hemp Cut Lime Mortar" (typically obtained from woven sacks or remnants of hemp cords, where the decomposed hemp rope residues were mixed with white plaster to enhance its resilience.)

From December 1968 to April 1974, I stayed in the countryside for more than 6 years and assisted a lot of peasant families in constructing their dwellings. During that time, local peasant families generally collaborated to build their homes without the involvement of professional construction teams. In those days, if you requested a temporary leave from village communal work, you would have your work-points deducted, resulting in a reduction in personal revenue allocated by the village administration; however, the homeowners would provide food for the helpers. Those who participated in the labor were referred to as "helpers". Since I had little knowledge about construction, I initially worked as a handyman performing tasks such as carrying bricks, pushing barrows, delivering raw materials, and even assisting in the kitchen duties like washing vegetables, pots and dishes, regardless of the task at hand. Instead of paying monetary compensation to helpers, homeowners expressed their gratitude by providing meals. In cases where families were truly impoverished and couldn't provide food, helpers still extended their assistance with compassion.

There was an old saying in the countryside, "Seeding in spring and harvesting in autumn". If anyone requested for a temporary leave from village communal work to help others build houses, when it was their turn to build their own houses in the future, generally other villagers would come to assist. Back then, I could not see any opportunity to leave the countryside and return to town. Therefore, as a relocated "educated youth" (rather than a local peasant), I had saved hundreds of catties of grains as potential gratitude for helpers in case I might build my house in the countryside later on.

Unexpectedly, I received a notice to continue my education in the city after the Spring Festival of 1974. Subsequently, I came to study at Beijing School of Civil Engineering and Architecture, which is now known as Beijing University of Civil Engineering and Architecture. After graduation, I remained with the college as a faculty member. Due in part to its nature as a construction school, the housing conditions for the staff were quite good, allowing married couples to be allocated an apartment. This was a fantastic welfare benefit provided by the work unit back in the 1970s and 1980s. Older people from that time can understand how challenge it was for families to be allocated an apartment.

After graduation, I got married. Since my wife was still serving in the army stationed at Xuancang City, Hubei Province, I became eligible for an assigned small room (12 square meters) provided by the school; it was the smaller room within a shared two-bedroom apartment. I lived there for about seven or eight years thereafter until the autumn of 1984 when my wife transitioned from military service to assume a new job in Beijing and our child started attending first grade at primary school. Subsequently, we relocated to an apartment measuring 22 square meters situated within a tube-shaped residential building on campus. However, this unit did not have its own kitchen, and I had to resort to cooking with gas tanks placed outside in the communal corridor—this arrangement being quite common during that period.

After the school was upgraded to a junior college, many senior teachers were hired. However, there were no available family dormitories. Therefore, bachelor dorms were transformed into family quarters for faculty members.

The College was originally designed with the help of Soviet Union experts during the early days after (the establishment of P. R. China). Teaching buildings, office buildings, student residences, faculty housing units and laboratory facilities all showcased Russian-style expansive roofs.

In 1988, my wife transferred to the Beijing Municipal Research Institute of Environmental Protection for work. The Institute provided us with a two-bedroom apartment measuring a construction area of 63 square meters on the condition that we return and re-allocate the 22-square meter apartment from the College to my colleague who was living off-campus. We resided in this 63-square meter apartment for ten years. By 1998, I was approaching retirement, our child had grown up, and both my wife's and my parents were aging, so we needed to relocate them to live with us. However, the 63-square meter apartment seemed too cramped for such a big family.

At that time, commercial houses could be purchased in the consumer market, while the original welfare housing distribution system was coming to an end. It was unrealistic to expect an improvement in housing conditions still allocated by the work unit. Faced with high housing prices and crowded old apartment, several of our colleagues initiated the idea of building houses in the countryside. As a result, we visited the countryside together to find a suitable location (the countryside was implementing a policy to support rural development at that time). After our hard work paid off a few months later, on the outskirts of Beijing, in the western mountain tail vein of Phoenix Mountain, we found a piece of a hillside wasteland. We began realizing our dreams of building new homes. Each colleague designed their own house style based on their economic ability.

I designed my own house and built a two-storey villa with my sister, consisting of two conjoined units accessed by separate stairways. Initially, I had planned to build a three-room bungalow that would provide ample space for restoring a long-lost pastoral lifestyle, offering respite from the crowded city we had endured for so many years and relaxing both body and mind. However, our desires grew a little bit wild. When each family presented their bungalow designs and discussed project costs, we suddenly realized that the average cost per square meter for a single-storey bungalow was much higher than that of a two-storey structure. As a result, we decided to modify our drawings into designs featuring two-storey villas.

After more than two years of construction and decoration, several small houses emerged on the barren hillside. They completely transformed our living environment over the past few decades, and all my colleagues and their families had been excited and happy about the new residences for several years.

However, despite being pastoral homes, the transportation conditions were very poor. Located in a mountainous area far away from the city, there was no bus line at all. This made our daily life and shopping extremely inconvenient; besides, heating was always a major concern during winter days due to the absence of a central heating system. Nevertheless, the ordeal of walking several kilometers to get home forced me to learn how to drive. In 2004, I learned to drive a car when I was already 55 years old. I regarded it as a blessing in disguise because even now at 74 years old, I still benefit from this skill—I can drive for shopping or seeing a doctor—and am able to manage everything by myself.

To make commuting to this rural home more convenient, I had to purchase a second-hand apartment in the nearby downhill village. The apartment, which measured 140 square meters in size, was spacious and well-maintained and had an elevator shared by two apartments. However, it lacked a complete property certificate from the authorities and was commonly referred to as "limited right property". Its primary advantage was the central heating system that required owners to pay heating fees on schedule regardless of whether they lived there or not, ensuring warm indoor temperature during winter. It consisted of three bedrooms, two bathrooms, and a spacious and uniform living and dining area measuring 45 square meters. With its double-balcony design, both additional areas were completely enclosed even during the construction stage. The south balcony, which was 1.5 meters wide with a floor-to-ceiling window measuring 4.5 meters in height, directly connected to the living room, creating a more spacious and bright space. The north balcony measured 1.3 meters in width and 4 meters in length connecting to the kitchen where we stored daily necessities and food items; especially during winter months when it served as an excellent natural large refrigerator and cold chamber.

The only drawback was the absence of windows in both bathrooms, not even a 50cm air vent. This shortcoming significantly reduced the value of the apartment as no matter how much attention had been paid to cleaning the bathrooms, they would inevitably emit sewage odor over the years, especially during hot summer days when the air pressure was low.

We lived there for another 10 years. Regardless its good maintenance, the location at the foot of the western mountain in the northern part of Beijing determined its inevitable chilly climate in winter. The corresponding outdoor temperature and annual average temperature were always 2-3 degrees lower than those of the urban districts in Beijing. Due to my back pain and rheumatism, the pain would always worsen in winter, resulting in my fear of coldness and preventing me from stepping out of the house during the five-month-long winter season. However, even though I enclosed myself at home for the entire winter season, it did not alleviate the pain. When I had to visit a doctor at a hospital, I happened to meet one of my teachers who advised me to go on a winter holiday to Hainan Province. He himself used to suffer from serious asthma but found excellent recuperation effects after spending two years of winter holidays in Hainan.

I have traveled to Sanya City, Hainan Province twice in the winter of 2003 and the spring of 2007. At that time, the housing prices in Hainan were quite cheap, but I never thought that I would live there for a long time in the future. In the winter of 2009, I set foot on my way to Sanya and stayed there for 62 days. During this trip,I completely fell in love with Sanya the treasure land, which is truly a paradise for retirees.

The winter in Hainan Province, especially in Sanya City, witnesses a wonderful resort filled with shady trees, coconut groves, beautiful bamboo trees, blooming flowers under the blue sky and white clouds. The distant sea beach is particularly charming. Singing and dancing are everywhere with various melodious musical instruments playing chords one after another. What a harmonious paradise!

The pace of reform has accelerated over time. On January 9th, 2010, the Central Government issued an order to develop Hainan Island into an international tourism destination. Taking advantage of the decline in property prices on Hainan Island, I purchased an 80-square-metre suite with elevators at Nanbin Farm in Yazhou town, Sanya City, which is only three kilometers away from the ancient town of Yazhou. Its southern exit connects to the highway around Hainan Province and is just one kilometer away from the Dongtian Scenic Spots (the Large/Small Fairyland Caves). Eight kilometers south lies Nanshan Temple, a world-famous Buddhist holy site where a 108-meter-tall white marble Buddha statue stands. This statue, known as "South China Sea Guanyin (Bodhisattva)" , is the tallest one in the world and has three heads facing three directions. Devout Buddhists who visit Hainan come here to worship this magnificent Buddha statue with sincere devotion.

Once again, this suite has limited property right; however, it is the most satisfactory one I have ever resided in. There is simply nothing to complain about. Whether it be the interior decoration style or the exterior environment, everywhere reflects consideration of human needs and care. Partially contributed to Sanya's unique geographical conditions, this city enjoys exceptional advantages in terms of residential environment.

Each room in the suite is equipped with a window, even if it faces the public corridor, ensuring that no kitchen or bathroom is enclosed without a window. As a result, its spatial permeability is particularly good. Upon entering the suite, there is a bathroom on the left with a window opening to the public corridor, while on the right there is a kitchen with a window facing westward. This suite enjoys smooth airflow from north to south, and maximizes interior layout without dead ends or wasted space. In between, there is a middle corridor with a withed of 2 meters that directly connects to the balcony. The balcony has been enclosed with bright French windows and connected to the living room, creating a spacious and comfortable area. Each of the two bedrooms has either a south-facing bay window or one facing westward, further enhancing interior spaciousness and brightness.

I am even more gratified by its public corridor layout design. The entire building is a triangular structure with two elevators in the middle, dividing the building into left and right halves, just like two hands of a person. Each floor holds twelve apartments, with six on each half. Six families share a stairway platform spanning over ten square meters in the middle, creating a courtyard-like space. Some Residents choose to set up tables on these platforms so that neighbors can play mahjong or cards together, or simply sit and chat while sipping a cup of tea. Due to the warm weather, neighbors always keep their doors open after waking up (only closing them when going out or sleeping at night). They can even casually chat with each other from their own apartments with doors open. Such platforms foster harmony among neighbors, making them feel like one big family. If someone needs onions or ginger for cooking, they can easily borrow some from their neighbors; similarly, if someone buys an abundance of vegetables and fruits at a good price, they are happy to treat their neighbors with them for free.

I have lived in northern China all my life. Regardless of whether two or three households resided on one floor, or whether there was an elevator, neighbors always closed doors and lived their own lives without interacting with each other. Even if there were occasions that required communication among the neighbors, they would knock on the door and speak at the gate with one person standing outside and the other inside; once conversations were finished, they would return home. Neighbors seldom invited each other into their homes. This kind of neighborhood seemed polite but particularly aloof. Sometimes, neighbors could hardly recognize each other when passing by in the public corridors.

I truly enjoy living in this apartment, probably because I have become afraid of loneliness as I grow older. I appreciate the courtyard-like layout, which creates a warm and intimate atmosphere. Another significant advantage of this residence, shared by many residential buildings in Hainan province, is that the ground floor does not accommodate any residents but instead serves as an expansive public hall for entertainment activities such as singing, playing chess or billiards, table tennis matches, exercising on fitness equipment, playing mahjong and more. The remarkable benefit is that we can take walks in the hall regardless of weather conditions, no matter whether it's sunny or pouring rain. This feature likely exists across residential communities throughout Hainan.

I have been reflecting briefly on my housing experience over the past few decades and have selected a few stages to record my feelings at that time. Honestly, I believe these instances adequately reflect the continuous improvement in living conditions and the promising prospect for happiness in the rapid development course of our country.

The Story of My Home

by Zhu Qingqing

"Wow, how splendid your home is!"

Ever since I started showcasing my home on Wechat Moments, I have frequently received such compliments. Everyone harbors their own dream of an impeccable residence, and for me, it took a complete span of 20 years to ultimately manifest my long-awaited dream home.

The Old House in Shuangyi County

I lived in a venerable abode nestled within Shuangyi County until I was eleven years old. This two-story building boasted a facade exhibiting the typical architectural style of traditional Anhui province. The east and west wings each hosted two chambers on the first and second floors respectively, with an atrium in the middle. The house was a wooden structure with its four walls constructed with rammed mud and without any cement. The house was comfortably warm in winter and cool in summer. However, even the faintest scurrying of a mouse could be heard distinctly through the creaking timber floors.

During difficult times, adult brothers of the family had to live cramped in the old house. As far as I could remember, there were always over a dozen people living together. Particularly during the "Peak Period", my family had to share the house with four additional households, including three uncles and one great uncle (the youngest brother of my grandpa). Halls and chambers were partitioned into cubicles to accommodate the large family. As a child, I perceived my home as exceptionally vibrant, especially during mealtimes when tantalizing aromas wafted back and forth from cooking ranges, arousing children's appetite to try various foods at each stove.

I really enjoyed this wonderful ambiance of having a big family.

As living conditions improved, relatives gradually moved out of the old house to their newly built homes, leaving many cubicles vacant. My father knocked down the partition walls and restored the original appearance of the old house—it turned out that the original central hall was actually very spacious. Gradually, I enjoyed sitting cozily in a chair and savoring leisure time there: Observing swallows constructing nests in spring, their chirping heralding the arrival of a new season; witnessing rain showers in front of the hall like a curtain during summer; listening to the wind rustling through the atrium and hall in autumn; and watching snowflakes gracefully descending one after another during winter. Even after being away from my hometown for many years, I still delighted in spending some time in the old house during holidays, even though it was no longer occupied. Thanks to the well-ventilated atrium, this abode remained impeccably preserved.

However, later on, the village began constructing a highway with a tunnel right through the mountain opposite the old abode. Since the old house was made of rammed mud years ago, the roadworks involving stone cannon operations shook it into a dangerous state, necessitating its demolition. On the day of the demolition, all family members went to the site, for we were deeply attached to the old adobe brimming with countless memories and emotions in our hearts. When the last wall was removed, it truly became an "old house" in our memories. A year later, my father constructed a new residence at the original location and specifically instructed the designer to include an atrium. However, amidst this new building made of steel and concrete materials, we could no longer discern that familiar vitality brought forth by our old abode. It was at that moment when I resolved to construct my ideal home in future endeavors.

Commercial Apartments

After the age of eleven, I had to accompany my father on his work transfers and constantly change our residences. Yes, those places could only be called "residences". Offices, workshops, warehouses, tool sheds—we experienced so many unexpected dwellings beyond imagination. Working outside, we could reside in any place as long as the family could live together. Naturally, they lacked the the essence of a true home. When my father eventually returned to work in the hometown county, we had the opportunity to buy a commercial apartment. At that time, it was a "big thing" to afford an apartment in the county town, especially for wage-earning employees like my father. We could hardly think about such a thing if there wasn't an employee housing benefits policy covering half of the house payment. The apartment spanned a modest 84 square meters with just one living room and two bedrooms. During this period, my sister and I attended school where most nights were spent in dormitories rather than at home while our parents worked at their workplaces during the day; therefore, home was primarily a place for sleeping at night and there wasn't much inconvenience caused by the humble size of the apartment. After marriage, I had my own apartment, which was also a commercial one in the county town, but I lost the novelty and contentment of residing in such a commercial apartment as I used to do. It became nothing more than a nest after a busy day.

The "Shuangqing Pavilion"

Returning to the countryside to build a dwelling had been a long-cherished aspiration brewing in my heart for many years. However, I could hardly make it for two reasons: firstly, we had to take care of our child who was attending school in town, leaving us with scarce time and energy to carry out this plan; secondly, the financial resources of the family were limited. Unexpectedly, a rather temporary accident compelled us to make a decision and take action. At that time, my parents-in-law lived in an old rural adobe erected during the 1990s. Although there seemed to be no major problems for their daily life, a prominent concern revolved around the toilet which was located outside the building. This meant that the elderly couple had to venture out at night merely for basic sanitary needs. Undoubtedly inconvenient as it was, what worried us most was the potential risk of stumbling accidents while we couldn't live together with them for daily care. Therefore, I asked a friend from design business to work out a solution of remodeling their house and adding a bathroom into it.

The initial site inspection of my friend turned out to be a shock: the entire building had been constructed using prefabricated concrete panels years ago, and there appeared to be small cracks everywhere. Initially, my parents-in-law were short of money, so they could only afford to build one room and gradually added more as they saved money over time. My friend mentioned that if we maintained the original layout, it might not collapse immediately; however, safety risks had already emerged. If we wanted modifications made, reinforcing the original walls would be imperative as a priority but ultimately cost would be as much as that of rebuilding the entire house. After carefully considering all these factors, we concluded that it was the right time to reconstruct the countryside house for the safety and convenience of my parents-in-law. At the same time, it presented an ideal opportunity for me to turn my dream into reality. Let's do it!

The advantages of having a designer friend became immediately apparent at this point. When I shared the idea of constructing a new countryside house with my friend, I conveyed specific requirements to him. For instance, we should not exceed the approved area by authorities; the public spaces should be spacious and bright; my parents-in-law should have relatively independent activity space; I needed a study room, and a yard should cater to both vegetable and flower cultivation. Eventually, he adopted a design concept of stacking blocks. Two square blocks were placed on the first floor with a triangle one above as the second floor, creating an integrated space encompassing a living hall, a dining room, a study and accommodation for the my parents-in-law on the first floor; meanwhile, two rooms adorned the second floor alongside an attached attic serving as both a leisure space and a bedroom. The south facade should be entirely composed of glass curtain walls, which not only enhanced the modern sense of the building but also met

My Countryside Home

A Corner View in the Courtyard

Dinner Table Set for Guests

Bedroom

our needs for an interior space with transparency and luminosity. We spent half a year completing the house construction, followed by nearly another year and a half devoted to meticulous decoration and courtyard embellishment—thus totaling two years. As eloquently expressed in poetic verses: "I possess a dwelling, facing glistening streams southward, adorned with resplendent blossoms of warm spring". At last came my dream home, and I named it "Shuangqing Pavilion" (*translator's note: Shuangqing in Chinese literally means clean and clear, which depicts pristine and fresh environment*).

It is only half an hour driving from the county town to my dream home, which is not far. However, due to busy work schedules, especially for my husband who always leaves for work early in the morning and returns late at night, we choose to drive to the countryside on weekends instead. My parents-in-law reside in this new villa and always keep it clean and tidy. My father-in-law is a skilled farmer who cultivates vegetables all year round. There is an abundance of fresh vegetables in the yard, so we don't need to buy any at the local markets; we are self-sufficient when it comes to vegetable supply. Furthermore, our garden thrives with dozens of flowers such as osmanthus, cherry blossoms, begonias, azaleas, maples, peonies, hydrangeas, roses, Shibazakura cherries, and Rosa banksiae. While my father-in-law takes care of the vegetable patch, I am responsible for tending to flowers and plants in the garden. A beautiful garden always requires meticulous care.

Every morning, as the first golden ray of sunlight ascends above the horizon, I am awaken by the melodious chirping of birds. Initially, I embark on an inspection stroll through the garden, watering plants, removing weeds, getting rid of pests, and fertilizing the soil. Afterwards, I enjoy a delectable breakfast in the garden while basking in the gentle sunshine before commencing my day in the idyllic countryside. Most mornings are spent preparing food in the kitchen; afternoons are dedicated to listening to music and writing articles in my study while sipping tea. Thanks to such a spacious house, I can invite friends over to enjoy parties with floral scenery and fragrant tea whenever we have free time—cherry blossoms in March, peonies in April, roses in May, hydrangeas in June... Though modestly sized, my garden flourishes with resplendent blooms throughout every season and bestows delightful surprises upon every nook and cranny.

Epilogue

During the compilation of *Kaleidoscope: Housing & Living (1949-2024)*, we hold a grateful heart for the thirty authors from all over the country and various walks of life who generously share their personal experiences and valuable insights in the articles. Their contributions enable us to catch a glimpse of the evolutionary trajectory of living environments since the establishment of the People's Republic of China, as well as perceive the unique impact that changing times have on each individual.

Each chapter of the anthology is a record of the authors' daily lives, delicately portraying their journey from childhood to adulthood, and eventually starting a family, even welcoming grandchildren. These captivating stories not only showcase the profound connection between individual destiny and national progress but also shed light on how changes in our living environment impact social structures, family relationships, and cultural heritages in various aspects. Through these authentic and vivid narratives, readers can fully immerse themselves in China's transformed social landscape over the past 75 years.

Here, I would like to express my heartfelt gratitude to Granny Baori Ledai, who has dedicated her life to improving living conditions and changing the natural environment in deserts alongside her fellow partners. She shares with us inspiring stories of how vast deserts have been transformed into oases. Sincere appreciation is also extended to Mr. Baoyu Pan from Longsheng Ethnic Autonomous County, who shares his invaluable experience in leading villagers towards poverty alleviation through tourism development while preserving landscape heritage.

Jose Manuel Ruiz Guerrero, a Spanish landscape designer residing in Beijing for many years, candidly expresses his genuine feelings about the changes occurring in China's urban environment. Furthermore, we are deeply grateful for the young teachers from China Agricultural University who volunteered on behalf of the China Youth Volunteer for Poverty Alleviation Relay Plan, as they share their brief yet colorful life experiences transitioning from the bustling capital city to remote frontier towns. Lastly, sincere acknowledgment is given to all the authors for conveying their cherished experiences and profound thoughts that truly breathe life and soul into this anthology.

In particular, I would like to express my sincere gratitude to the Chinese Society for Sustainable Development, China Architecture Design and Research Group, National Engineering Research Center for Human Settlements, as well as the leaders and colleagues from the innovation demonstration zones of National Sustainable Development Agenda (including Guilin City, Linchuan City, Ordos City, Zaozhuang City, Hainan Tibetan Autonomous Prefecture, Chengde City, Shenzhen City, Huzhou City, and Chenzhou City). They have provided us with tremendous support in compiling this series. It is their recognition of the long-term efforts of the Special Committee for Human Settlements that has enabled us to present *Kaleidoscope: Housing & Living (1949-2024)* as a tribute to the 75th anniversary of the People's Republic of China.

As members of the Chinese Society for Sustainable Development, we are fully aware of our responsibilities and missions. Since its establishment, we have always adhered to a people-oriented philosophy and placed great emphasis on continuously improving and developing human settlements. We actively engage in domestic and international communication and cooperation, offering Chinese wisdom and experience to promote global sustainable development goals. The English version of our latest series, *Kaleidoscope: Housing & Living (1949-2024)*, is scheduled for release at the UN World Urban Forum in Cairo, Egypt's capital, in November 2024. Our aim is to proactively share with the world the stories of China's sustainable development in human settlements.

Once again, I would like to express my heartfelt appreciation to our editorial team and all those who have supported and assisted us. Their unwavering support and encouragement have been invaluable in helping us complete this significant collection. We sincerely hope that this anthology will raise awareness among a wider audience about the importance of our living environment for both individuals and society as a whole, inspiring more people to actively participate in activities aimed at enhancing our living environment and promoting sustainable development. Let's join hands together to create a more harmonious, livable and beautiful homeland.

Xiaotong Zhang
Executive Director of Chinese Society for Sustainable Development
Secretary General of the Special Committee for Human Settlements
September 2024